T0224519

Fortschritte Naturstofftechnik

Reihe herausgegeben von
T. Herlitzius, Dresden, Deutschland

Die Publikationen dieser Reihe dokumentieren die wissenschaftlichen Arbeiten des Instituts für Naturstofftechnik, um Maschinen und Verfahren zur Versorgung der ständig wachsenden Bevölkerung der Erde mit Nahrung und Energie zu entwickeln. Ein besonderer Schwerpunkt liegt auf dem immer wichtiger werdenden Aspekt der Nachhaltigkeit sowie auf der Entwicklung und Verbesserung geschlossener Stoffkreisläufe.

In Dissertationen und Konferenzberichten werden die wissenschaftlich-ingenieurmäßigen Analysen und Lösungen von der Grundlagenforschung bis zum Praxistransfer in folgenden Schwerpunkten dargestellt:

- Nachhaltige Gestaltung der Agrarproduktion
- Produktion gesunder und sicherer Lebensmittel
- Industrielle Nutzung nachwachsender Rohstoffe
- Entwicklung von Energieträgern auf Basis von Biomasse

This series documents the Institute of Natural Product Technology's work to develop machinery and processes to supply the world's continuously growing population with food and energy. It particularly focuses on the increasingly important aspect of sustainability and the development and improvement of closed material cycles.

Theses and conference reports document engineering analyses and solutions from basic research to practical transfer along the following focal topics:

- Sustainability of agricultural production
- Production of healthy and safe food
- Industrial use of renewable raw materials
- Development of energy sources based on biomass

More information about this series at http://www.springer.com/series/16065

Benjamin Seiferth

Development of a system for selective pasture care by an autonomous mobile machine

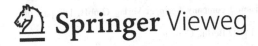 Springer Vieweg

Benjamin Seiferth
Memmingen, Germany

A dissertation submittet to the Faculty of Mechanical Science and Engineering at the Technische Universität Dresden for the degree of Doctor of Engineering (Dr.-Ing.)
by
Dipl.-Ing. Benjamin Seiferth

Reviewer:
Prof. Dr.-Ing. habil. Thomas Herlitzius, TU Dresden
Prof. Dr. rer. nat. Arno Ruckelshausen, HS Osnabrück

Day of Submission: 13.06.2018
Day of Defence: 13.01.2020

Chairman of the Promotion Commission: Prof. Dr.-Ing. André Wagenführ, TU Dresden

ISSN 2524-3365   ISSN 2524-3373   (electronic)
Fortschritte Naturstofftechnik
ISBN 978-3-662-61654-3   ISBN 978-3-662-61655-0   (eBook)
https://doi.org/10.1007/978-3-662-61655-0

This Springer Vieweg imprint is published by the registered company Springer-Verlag GmbH, DE part of Springer Nature.
The registered company address is: Heidelberger Platz 3, 14197 Berlin, Germany

Preface

The present thesis was created in the frame of the European research project „i-LEED" during my time at the Institute for Agricultural Engineering and Animal Husbandry (ILT), a department of the Bavarian State Research Center for Agriculture (LfL), between 2014 and 2017.

The i-LEED project "Advanced cattle feeding on pasture through innovative pasture management" was funded within the 2nd ICT-AGRI ERA-NET and financially supported by a grant from NCPs: the German Federal Ministry of Food, Agriculture and Consumer Protection (BMELV) through the Federal Office for Agriculture and Food (BLE) grant number 2812ERA058 and 2812ERA059, the French National Research Agency (ANR) grant number 12-ICTA-0002-01 and 12-ICTA-0002-02 and the Scientific and Technological Research Council of Turkey (TUBITAK) grant number 112 O 464.

First of all, I would like to thank my doctoral supervisor Prof. Dr.-Ing. habil. Thomas Herlitzius who drew my attention to the job of the project and thus the work initiated. I am very grateful for his advice and regular motivation during the past years. I am thanking Prof. Dr. rer. nat. Arno Ruckelshausen for his role as second supervisor.

Especially, I would like to thank my colleagues at the ILT under the leadership of Dr. agr. Georg Wendl and Dr. agr. Markus Demmel. I owe special thanks to Stefan Thurner and Dr.-Ing. Georg Fröhlich for leading and organising the project. In addition, I am also very grateful for the uncomplicated assistance by Michael Wildgruber with his workshop team, Heiner Link, Thomas Kammerloher and Robert Weinfurtner. They contributed to this work with much practical support. In this context I would like to mention Gabriele Ostermeier who always clarified organisational matters. Moreover I would also like to express thanks to the French colleagues within the project for the great hospitality at the National Research Institute of Science and Technology for Environment and Agriculture (IRSTEA) in Clermont-Ferrand under the leadership of Michel Berducat. Special thanks goes to Dr.-Ing. Christophe Cariou and Dr.-Ing Cedric Tessier who contributed to this work with much expertise in the domain of vehicle guidance. I really enjoyed my three-month stay in the Auvergne.

Besides, I would like to thank my parents, my brother, grandparents and friends for the moral and educational support. Finally, the greatest thanks for his help goes to God, the father of Jesus Christ, who is the creator of the heaven and the earth and thus the best engineer ever.

The book is intended to show engineers and researchers who work on agricultural machinery how new challenges of digitization in mechatronic machine systems can be implemented in agricultural engineering by new ideas but also by the meaningful integration of known technologies and components. This is exemplified by the development of a machine for automated pasture care.

Memmingen, February 2020   *Benjamin Seiferth*

Index of Contents

List of Abbreviations

AFS	Automatic Feeding System
AMS	Automatic Milking System
CAN	Controller Area Network
COB-ID	Communication Object Identification
DGPS	Differential Global Positioning System
GSM	Global System for Mobile Communications
GPRS	General Packet Radio Service
GPS	Global Positioning System
ILT	Institute for Agricultural Engineering and Animal Husbandry
ISO	International Organization for Standardization
IVT	Information and communications technology
LADAR	Laser Detection and Ranging
LiDAR	Light detection and Ranging
LLC	Low Level Controller
OMC	Original Machine Control
PDO	Process Data Objects
PRT	Pulse Ranging Technology
RF	Radio Frequency
RFID	Radio Frequency Identification
RTK	Real Time Kinematic
UHF	Ultra High Frequency
VRE	Value of Received Energy
WLAN	Wireless Local Area Network

Index of Symbols

Symbol	Unit	Description
A, B, C	-	points of joints
$\dot{A}_{cattle}$	ha/h	area performance of herd of cattle
$\dot{A}_{machine}$	ha/h	area performance of the machine
A_p	ha	paddock area
A_{sh}	mm²	area of laser shadow
a_l, b_l, c_l	mm	lever lengths of steering kinematics
a_{spots}	mm	distance between two un-grazed spots in direction of travel
b	mm	calculated necessary ground clearance
b_e	g/kWh	specific fuel consumption
c	mm	curvature of the vehicle trajectory
c_{auto}	$	costs caused by automated pasture care
c_{conv}	$	costs caused by conventional pasture care
$cmd_{steering}$	-	command value for driving
D	m	diameter of circles driven by the vehicle
d_s	mm	diameter of screw reference circle
d_t	mm	measured travelled distance
d_w	mm	wheel diameter
e	€	equivalent of forage
f	Hz	frequency of laser head rotation
f(c)	-	polynomial functions with variable c
j	-	number of the laser beam relating to the centre one
L	mm	wheel base of the vehicle

v_t m/s   transport speed

v_x m/s   vehicle speed in x-direction in relation to the vehicle coordinate system

w mm   working width

w_{scan} mm   width of scanning zone

w_t mm   track gauge of the vehicle

x_B m   x position of detected soil profile

x_{B_f} m   x position of filtered soil profile

x_0, y_0, z_0 m   global position of the vehicle

x_{0i}, y_{0i} mm   points of planned trajectory

z_{alt} mm   altitude of the vehicle

z_B m   altitude of the soil contact point B / z position of detected soil profile

z_{B_f} m   z position of filtered soil profile

z_g mm   height of grass

$\bar{z}_{gn}$ mm   grass height average in each section of scanning zone

z_{laser} mm   attaching height of the 2D laser scanner

z_{max} mm   maximum mulcher position

z_t mm   set threshold for grass height

z_{v_a} mm   position of GPS antenna onboard the vehicle

1   Importance of pasture grassland and its careful maintenance

1.1   Worldwide trends of the livestock farming sector

In the face of an increasing world population and the accompanying higher demand for food products, agricultural land is becoming increasingly valuable. This trend is especially applicable to worldwide permanent grassland areas, which is an important resource for fodder production for livestock, in particular cattle. As in the past, the food consumption of bovine meat, milk and dairy products will probably continue to rise [1]. In the mid-1960s the average worldwide consumption of meat per capita amounted to 24.2 kg. Since that time it increased continuously in over the last decades (see Figure 1). In 2030 it is expected to exceed 45 kg. The same applies to milk and dairy products. The average worldwide consumption of milk and dairy products per capita was 74 kg in the mid-1960s. By 2030 it is expected to increase to 90 kg. [2]

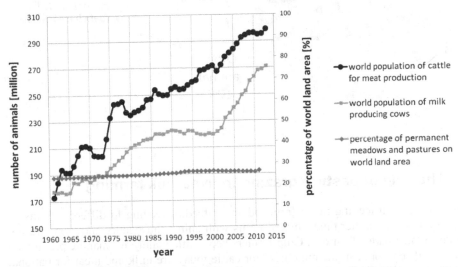

Figure 1:   Worldwide development of cattle and dairy cow populations compared to the trend of global grasslands to date (according to data from [3], [4], [5], [6], [7], [8])

This increase in demand for bovine meat and milk products has resulted in an ever growing livestock farming sector. Figure 1 shows the trends of the animal population and grasslands in the world in the last 50 years. The worldwide number of cattle for meat production has increased by approximately 75 %, from 170 to 300 million animals between 1960 and 2013 ([4], [8]). Similarly the population of milk producing cows in the world has risen by 50 % from 180 to 270 million animals in the same period ([3],[5]). The majority of this increase in cattle and dairy cows in this period was found in Africa, America and Asia. In Oceania the number of cattle and dairy cows has nearly stagnated, and in Europe it has even decreased. Nevertheless, the area of permanent

© The Editor(s) (if applicable) and The Author(s), under exclusive license
to Springer-Verlag GmbH, DE, part of Springer Nature 2020
B. Seiferth, Development of a system for selective pasture care by
an autonomous mobile machine, Fortschritte Naturstofftechnik,
https://doi.org/10.1007/978-3-662-61655-0_1

meadows and pastures in the world has remained nearly constant. It has only increased by 2.6 % during the aforementioned period [6]. One consequence of this is a more intensive use of the area or housing systems by using concentrated feed. Due to these trends the available area of grassland for fodder production per animal has been reduced worldwide. Permanent meadows and pastures have become more and more valuable for fodder production. Although livestock numbers have decreased in European countries in the past two decades, the milk and beef sector had a share of the output value in the EU-28's agricultural industry of more than 20 % in 2015 [9] (see Figure 2). Farmers are under increasing pressure to produce more cost-effectively to be able to compete with other sectors, also with respect to biomass production.

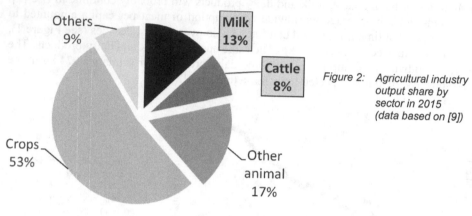

Figure 2:   Agricultural industry output share by sector in 2015 (data based on [9])

1.2   The role of pasture grazing in livestock farming

Methods of animal feeding range from indoor or feedlot feeding to different forms of grazing systems. In earlier times, mixed farms with animal husbandry and crop farming came up in most parts of Europe. Crop farmers kept cattle mainly for manure production in barns [10]. By contrast, the main uses for cattle today are milk and meat for national and international markets and manure is now regarded as a by-product [10]. In 2010 the number of farms with grazing livestock within the EU-27 amounted to 58 % [11]. In Germany this figure was only 44 % [12].

Disadvantages of pasture grazing include challenges in managing large herds. The performance drops without supplement feeding in the high-performance range of dairy cows and another disadvantage is the widely varying quality and quantity of available forage during the season. Yet current discussions about greenhouse gas emissions, climate protection, animal welfare and biodiversity have brought more attention to permanent grasslands again [13]. Indeed, grazing is the natural way and can be a highly efficient method of animal feeding [10]. Some experts are also talking about the "low-cost method" in connection with milk production based on grazed grasslands. Conversely livestock housing is called the "high performance method" [13]. Lower costs are one advantage of pasture grazing, especially the cost of the staple diet. There are also no losses by forage conservation. The workload of feeding and cleaning is reduced at least seasonally. Lower capacities for silage preparation and storage, as well as for storage of manure, are needed compared to all year indoor feeding. [13]

Public opinion and organizations promote naturalness and sustainability and regard grazing cattle, sheep and horses as part of the landscape [10]. Surveys among consumers confirm this attitude [14]. The majority of consumers, at least in Germany, think that pasture grazing improves the animal welfare. In fact, several surveys have proved that pasture grazing promotes animal health [15]. Fewer problems with limbs and claws occur, including an improved cleanliness of the animals, if the pasture is well maintained [13]. Thus, the livestock farming sector has public commitment to improve animal protection in practice. Existing conflicts between animal welfare and economy must be recognised and reduced [13]. Animal protection is becoming a more important aspect when considering housing systems, as consumers are even willing to pay more for milk from grazing cows. Some dairy plants have recognised this willingness and offer meadow-grazed milk [13].

The overall challenge in fodder production or secondarily milk and meat production is to move the centre of focus away from the economic aspects and towards a better balance with ecological and societal aspects. The goal is to create economic value without compromising the impact on ecological and societal aspects as much as it is done today [16]. Figure 3 illustrates this described aim. The use of modern information and communication technology (ICT) in agricultural engineering could potentially contribute to achieve this goal.

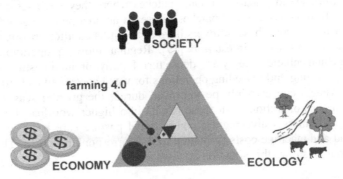

Figure 3:   Illustration of the balance between different interests achieved by farming 4.0
            (based on [16])

1.3   The role of pasture maintenance

A good turf of pasture is the basis of high grazing performance and an efficient grazing farm. Insufficient pasture maintenance can lead to losses of pasture forage, in terms of both quantity and quality. Incorrectly set mowers or driving on the paddock area under wet conditions can cause damage to the turf [17]. Additionally, particular zones with much traffic volume of cattle on pasture, e.g. the water trough, are usually affected by damages by the footsteps of cattle (see Figure 4). Reseeding of zones is necessary to improve the turf density.

Figure 4:   Damages to the sward left by cattle   Figure 5:   Typical spots of leftovers after grazing
             hoofs   on pasture areas

Furthermore, cattle avoid grazing on some patches where they have set urine or feces (see Figure 5). In worst case, those patches can reach an area percentage of up to 10 to 20 % [18]. For this reason, these leftovers have to be mulched after grazing to prevent a further expansion and weed infestation [19]. Regular mowing operations eliminate outdated forage and enhance a leafy and dense turf. Finally, optimal pasture maintenance includes both mulching and reseeding operations for weed control and sward renovation. Pasture effectiveness can certainly be improved during the grazing season by regular pasture maintenance operations, but these results in a higher workload for the farmer. For this reason, farmers waive regular conventional pasture care because they expect more workload and machine costs compared to the corresponding profit. The following diagram in Figure 6 shows this situation qualitatively.

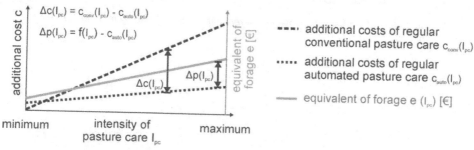

Figure 6:   Qualitative illustration of the profitability of pasture care

The additional costs as well as the equivalent of forage in € increase with the intensity of pasture care. Yet automated pasture care is only worthwhile if the associated additional costs are lower than conventional pasture care ($\Delta c > 0$) and the related equivalent of forage so that there is a profit ($\Delta p > 0$).

2   State of the art on grazing farms and the trend of automation in agriculture

Similar to the advancements being made in industry, the current trend towards automation and data exchange is also present in agriculture. This includes cyber-physical systems, Big Data and Human Machine Interface technologies [16]. This trend is shown by developments at research institutes and also by new prototypes and products from companies. The focus is among others on automated intelligent small machines (swarm technology [16], [20]). The idea is not to fully replace the human as operator, but to separate the operation and the operating process locally or rather temporally. Until now pasture maintenance has been performed mechanically and manually. In the communal sector there are some remote controlled commercial products, which show the first steps towards separate operation and operator (see paragraph 2.2). By contrast, automation in livestock housing is quite advanced. Paragraph 2.3 shows examples. The GPS-based automatic guidance system has already been established in crop production. Other developed automated machines are rather prototypes and not yet commercial products (see paragraph 2.4).

2.1   Manual and mechanised technique of pasture maintenance

Until now the mowing operations of un-grazed spots have been done selectively but only manually, by tools like scythes or small motor mowers (see Figure 7). Selective seeding is done by hand or small tools. Big machinery is usually only used for mulching, grooming or reseeding the whole area once or twice a season or there is no pasture maintenance at all.

One well-tried tool for mowing is the scythe. It is especially lightweight and easy to transport in difficult terrain. A disadvantage, however, is the sensitivity of the scythe blade. Maintaining a brush cutter is easier and its sensitivity is lower. Both tools have low area efficiency and are only used on small pasture farms or on pastures with difficult terrain, e.g. alpine pastures. Further machines for smaller areas include single-axle, hand-guided machines with a disc mower, flail mower or cutter bar. The operator walks behind the machine and has to steer it with muscular strength, which can be very strenuous [21].

Figure 7:   Hand-held brush cutter [22] (left), single axle mower[21] (centre), hand-operated tool for seeding operations (so-called "clover violin") [23] (right)

Larger, tractor driven machines are usually only used for mulching, grooming or reseeding the whole area once or twice a season or, as mentioned, there was no pasture maintenance at all. Generally there are attached mowers with rotational and oscillating tools, like cutter bars (see Figure 8). The rotational tools can be subdivided into tools with two types of rotary axes: horizontal and vertical. Flail mowers have a horizontal rotating drum, transverse to the direction of travel, which is equipped with so-called flails of various designs depending on the operating conditions. They are hinge-mounted and align radially by centrifugal force. [24]

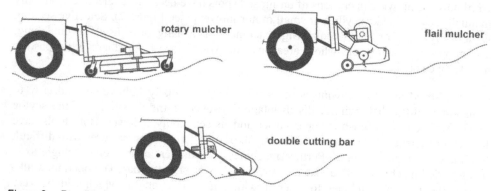

Figure 8:   Possible types of machines for removing un-grazed spots on pasture

Mowers with a vertical rotary axis are simply called rotary mowers. Rotary mowers consist of one to four hubs across the working width. Each hub has cutting tools, like blades or sickles. The cutting tools can be swivel-mounted or fixed [25]. The height guidance of machines with rotational tools can be achieved by using skids, wheels or rollers. Cutter bars are mostly supported by lateral skids. Despite low costs of acquisition cutter bars have not been able to establish themselves in agriculture. In the landscape maintenance sector they are often used for single-axle, hand-guided machines, as explained above. There are only a few manufacturers of tractor-attached cutter bars. These machines are often so-called mulching bars and double cutting bars. Finger bars with permanently installed steel fingers are rarely used today because they tend easily to block up with forage material. An optimisation of the finger bar, is the mulching bar which uses fixed triangular blades. A further improvement is the double cutting bar. It consists of two rows of blades which move in opposite directions. Therefore, higher operating speeds are possible. [24] [22]

In principle, there are two methods for the sowing seeds on grassland or pasture. In the first, the seeds are cast onto the ground surface, and in the second, seeds are introduced into the soil. The seeding operations are mostly combined with other operation processes. If there are fewer or smaller missing areas of the sward, seeds can be cast by hand. Similar to the hand seeding process, but more effective, is the use of a fertilizer or slug pellets spreaders, which can be attached directly at a tractor or other attachments, for example the pasture harrow. A further possibility for a non-tillage seeding operation is the use of a seed driller without seeding pipes. Usually seed drills introduce seeds into the ground. They differ in their tillage tools for the preparation of the seedbed. Discs or coulters are typical tools of seed drills. [26]

Figure 9:   Tractor-driven mulcher [21] (left), pasture seeder harrow combination [23] (right)

2.2   Remote controlled technique to maintain green landscape areas

Remote controlled machines are more and more frequently used in the municipal sector, especially to maintain impassable green areas. The operator is not at the machine. Thus, he can control the machine from a safe place and the machine can work in difficult terrain. He is not directly involved in the process, which would otherwise be outside his comfort zone, e.g. due to noise, vibrations and dust. The machine is remotely controlled usually by a radio interface. Most of these machines are equipped with a mulcher, which requires a combustion engine because of the high power requirement (10 to 20 kW per metre of working width) and the related long operating time. The remote controlled machines have a wheel or crawler chassis. Figure 10 gives an overview of the commercial machines with different installed power. The smallest remote controlled mowers start with around 5 kW of installed power and an operating width of 600 mm. The installed power of these machines increases with the operational width. Smaller ones are often equipped with a rotary mower and are used for lawn areas. Larger machines have a flail mulcher which can be used on rough grasslands.

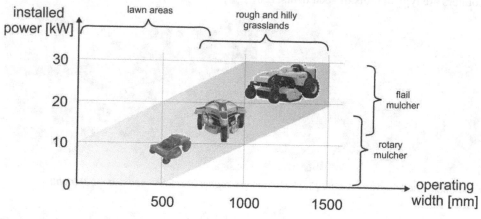

Figure 10: Overview of remote controlled mowers with regard to installed power and operating width
(sources of pictures: [27], [28], [29])

2.3   Automation in livestock farming

Automation in agriculture, especially in livestock housing farms has advanced greatly in recent years. The availability of high computing, storage capacities, high-performance environmental sensors and cameras enable a development of intelligent systems for automation on stationary and mobile systems. In livestock housing systems the degree of automation has increased in the recent years. Automatic milking systems (AMS) are used more and more frequently. The worldwide number of farms which use such a system has increased in the last ten years. At the end of the year 2010 the estimated number of farms which used an AMS was 10,000 [30].

Automatic feeding systems (AFS) are also used on farms. In 2013 approximately 1000 European farms applied AFS [31]. AFS range from stationary systems such as conveyor belts to mobile system such as self-propelled or rail-guided feeder mixer wagons. The systems also differ in the level of automation depending on the task required, e.g. removing, transport, mixing, distributing and pushing the feed. The mobile robot systems which are available use odometers, ultrasonic sensors, guide wires or radio-frequency identification (RFID) tags for orientation. The forage is regularly loaded out of a container, mixed, and placed at the feeding fence. Additional mechanical devices of the system push the forage periodically back towards the feeding fence [32].

Figure 11: Lely Juno feed pusher robot [33] (left); DeLaval robot scraper [34] (right)

Cleaning robots in barns also have established. The manure is removed by a robot scraper or a mobile barn cleaner. At regular intervals the robot starts automatically and follows a pre-programmed path. The orientation works just like the mobile AFS robots (see above).

Automation of tasks in grazing systems is rather seldom. Aside from some research projects and niche products (e.g. [35]), there is no commercial widespread automated system for grazing. In this context the automatic grazing system from manufacturer Lely must be mentioned. It consists of two mobile robots with a solar panel, which provides power for each battery. A wire is stretched between the two robots and thus, they allocate the pasture area by means of pre-programmed time and number of metres the fence should move per day. The system can also be used for driving the cows back, e.g. for milking [36].The use of stationary AMS in combination with pasture grazing is possible [37], but is not yet common. There are also approaches for mobile AMS, especially for grazing systems [38], but there is no commercial product on the market. Recently the University of Sydney reported on a robot project, in which a robot for monitoring and herding cattle on pastures is to be developed [39].

Figure 12: Approach of a mobile AMS for the pasture [38] (left); Lely voyager robot for automated fence moving [36] (right)

2.4   Automation of mobile agricultural machinery in crop production

The automation of machines has also been established in other sectors of agricultural engineering. Tillage, harvest and maintenance operations are affected by the progression of automation. While commercial products are semi-automated, such as the height guidance of certain implements or steering systems, well known manufacturers are also working on autonomous solutions in terms of precision farming. For example, the manufacturer of agricultural machinery Fendt has started the research project "MARS" (now called "XAVER") which stands for Mobile Agricultural Robot Swarms. A small and lightweight electrical driven robot with a precision seeding unit is being developed which will be a part of a swarm consisting of many identical robots [40] [41]. The manufacturer Kongskilde has presented a machine called "Vibro Crop Robotti" back in 2013. It is a robotic platform consisting of two individual electrical belt-driven units, between which different tools can be equipped. The primary tool is a cultivator for cutting roots of weed in row crops [42]. The companies Amazone and Bosch have developed a robotic platform, called BoniRob, for both screening and machining tasks for row crops in cooperation with the university of applied sciences Osnabrück. The platform with four independently driven wheels can be equipped with different application modules [42]. The following figures show the machines of the afore mentioned development projects of well known manufacturers in agricultural engineering.

Figure 13: Fendt XAVER [40] (left); Kongskilde Robotti [43] (centre); BoniRob [44] (right)

One reason for this trend towards automation is the enhancement of sensors and computing power over the years [45]. Prototypes have shown that machine operations in agriculture can be automated with the aid of different sensors. The guidance of vehicles in free and structured fields, e.g. plantations, as well as the obstacle detection are important operations in this context [46]. Other implement operations are intelligently controlled based on sensor measurements in terms of precision farming [20]. Risks have to be minimized and collisions with humans, animals or other machines must be prevented before such machines are introduced onto the market. Thus, obstacle detection and collision avoidance functions are essential for automated machines [47]. Legal issues have to be clarified first. There are already drafts of the International Organization for Standardization (ISO) to define international standards for highly automated machines [48], but they may not be referred to as an international standard. Global Positioning Systems, in addition to laser, visual, radar and ultrasonic sensors are the main methods used for localisation and navigation as well as intelligent implement control and obstacle detection (sensor fusion).

3   Added value of pasture grazing by automated pasture care

Robots have already been introduced as commercially available milking or scraper robots in animal housing on dairy farms (see paragraph 2.3). As described in chapter 2, the level of automation has increased in agriculture in recent years, especially in livestock farming, but there are other sectors which have potential in regard to automation. This includes outdoor sectors like pasture grazing, in particular where the degree of automation is still quite low.

Generally, there are still no commercial small autonomous machines for outdoor agricultural areas represented by leading manufacturers of the agricultural machinery. The obvious current trend is towards the development of stronger, faster, wider and bigger machines [49]. Contrarily, scientists regard smart and small machines (swarm) as a possible concept for cultivating agricultural land in future [16] [20]. This concept would be also an opportunity to optimise pasture grazing. Currently pasture maintenance operations are often carried out once a year with either high area productivity on the whole area, or no pasture maintenance at all. The idea of precision farming is "doing the right thing in the right place at the right time" [20]. Mulching of un-grazed leftover spots and reseeding of damages by footprints after grazing are usually not necessary for the whole area. Therefore, selective pasture maintenance is completely sufficient and resource-friendly. Extensive contamination of forage by mowing the whole area is even avoided [50]. Damages to the fauna can be reduced and, thus, biodiversity is conserved and enhanced [51]. However, positive aspects from an economical point of view the nature conservation standpoint aside can be seen. According to paragraph 1.3 regular pasture care increases the quality and quantity of herbage and also the efficiency of feeding animals. Moreover, a lower energy requirement by carrying out selective mowing operations is expected, as well as lower wear of tools. As described in 1.3, a profitability of the new concept must be the objective. According to this background the task of this thesis is to develop and evaluate a system for selective pasture care by an automated mobile machine based on conventional technology for mowing and seeding operations (see paragraph 2.1). To enable automatic and selective pasture maintenance, relevant spots must first be detected. The following Figure 14 illustrates the definition of the task with the objective to automate pasture care by the fusion of previously mentioned technologies, including the accompanying electrification.

B. Seiferth, *Development of a system for selective pasture care by an autonomous mobile machine*, Fortschritte Naturstofftechnik, https://doi.org/10.1007/978-3-662-61655-0_3

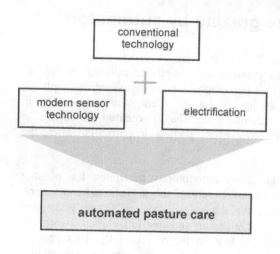

Figure 14: Fusion of conventional and
enabling technologies for
automation of pasture care
operations

4   Potential technologies for the automation of pasture care

4.1   Global Positioning System (GPS)-based navigation

The enhancement of the Differential Global Positioning System (DGPS) in particular has contributed to a trend of automated machine guidance [46]. The Real-Time Kinematic Global Positioning System (GPS-RTK) and the Differential Global Positioning System (DGPS) are an enhancement of the Global Positioning System (GPS) and use additional correction signals from a network of fix ground-based reference stations. In principle, the reference receiver determines the difference between the computed and measured range values (correction) to broadcast the difference between the positions indicated by the GPS satellite systems and the known fixed positions to the machine [52]. These correction signals can be received by mobile radio (e.g. GSM, GPRS). If correction signals are not available from a provider, a local reference station at the farmyard can be installed and linked by a radio interface to the rover. Thus, an accuracy to within 2 cm [53] allows a precise navigation of the machines on free area fields and the operations can be carried out along a pre-determined path. Areas can be machine operated by following trajectories which fully cover the area in distances corresponding to the working width of the implement. The most common disadvantages of using GPS include the loss of communication, if the line-of-sight with satellites is obstructed, e.g. by trees, multi-path issues, and interferences from other RF sources [54].

4.2   Laser-based automation

In many robotics projects at research institutes laser scanners for navigation and environment recognition of their prototypes are in use. N. Shalal et al. [46] even call the laser scanner the most popular device in outdoor applications. A high resolution and a large field of view as well as the reliability and robustness at different weather and ambient illumination conditions are the reasons for its frequent use [46]. 2D laser scanners are often used. Their method is sometimes also known as light detection and ranging (LiDAR) [55] or Laser detection and ranging (LaDAR) [56]. 2D laser scanners consist of a rotating sensor head where the light source and the light receiver are located. Thus, the relative distance of objects is determined by measuring the time of flight of laser pulses in certain angular distances $\Delta\phi$. A laser scanner, for example, is used among others for navigating a service-robot through rows of vineyards and orchards within the project "elWObot" [57]. P. Thanpattranon et al. [58] used a laser scanner as a single sensor to detect objects and navigate a tractor through the rows of a tree plantation (see Figure 15). Another example of laser data based application is the development of advanced sprayer controls, which depend on the size and density of the tree canopy in plantations [56] (see Fig. 13). Obstacle detection based on laser scanner data is common today [59] [60]. Some commercially available 2D laser scanners also output other values besides the distance values. These values without any physical interpretation are called "intensity" values [61] and characterise the received energy of the reflected laser beam depending on the surface properties and the distance from sensor to the object. The strength of the backscattered light beam is influenced by the specular and diffuse reflection behavior, absorption and transmission. Thus, a calibration can enable the conversion of these intensity values into parameters related to a specific object surface [61], which can be localised in that way.

Figure 15: LiDAR navigated tractor [58] (right); automated nozzle control of a sprayer based on ultrasonic sensors[56] (left)

4.3 Vision-based automation

Vision-based guidance is widely used in the automatic guidance of prototype machines, because of the good price-performance ratio and their capability to provide huge information [46]. For example, Billingsley and Schoenfisch [62] were engaged with the development of vision-guided tractors. A machine steering system for agricultural combines was developed at the University of Illinois to navigate the combine along the cut edge [63]. Cameras are also often focused on the scientific research for obstacle detection in agricultural machines [57] [64]. One decisive disadvantage of using vision sensors is the influence of varying ambient lightening conditions, especially in outdoor environments [46].

4.4 Automation based on other sensors

Ultrasonic sensors are used for the navigation of vehicles, too. An automatic guidance system for tractors in fruit farming based on an ultrasonic sensor measurement combined with a laser was presented by [65]. There are also existing commercial sprayers which use ultrasonic sensors to activate nozzles based on the presence of trees [56]. But the application of ultrasonic sensors is restricted in some ways because a proper reflection of the ultrasonic echoes requires the objects to be perpendicular to the sensor [46]. They have a low spatial resolution [56].

4.5   Benchmark of different technologies

In summary, Table 1 lists the main technical advantages and disadvantages of the described sensor technologies.

Table 1: main technical advantages and disadvantages of different potential technology ([46] [50] [51] [53])

Technology	Advantage	Disadvantage
Global Positioning System	• accuracy within 2 cm	• loss of communication in cases of line of sight with satellites obstruction and interferences from other RF sources
Laser scanner	• high resolution • large field of view • reliability and robustness • delivery of more information beside distances	• mechanically moving parts • relatively large assembly space
Vision-based sensors	• capability to provide huge information • large field of view	• influence of varying ambient lightening conditions
Ultrasonic sensor	• high robustness	• objects have to be perpendicular to the sensor for the ultrasonic echoes to be reflected back properly. • low spatial resolution

5 Requirements for automation of pasture care by a mobile machine

5.1 Technique of grazing management

There are different procedures of grazing management, which is applied depending on the intensity of grazing, animal movement and layout of the available pasture areas. The different grazing methods are important for the definition of constraints and requirements of operating machinery. Especially the logistics process of the machine has consequences on a lot of parameters of automated pasture care. According to grazing intensity there are two types of grazing: extensive and intensive grazing (see Figure 16). [26]

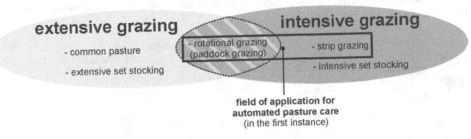

Figure 16: Classification of grazing methods according grazing intensity and field of application for automated pasture maintenance (based on [26])

Extensive grazing includes the common pasture and the permanent pasture. Common pasture means overgrazing the area one-time, what is for example appropriate for nomadic shepherds. A further extensive method is continuous stocking on permanent pasture. On permanent pasture the animals always have access to a specific unit of land, which is not subdivided in smaller parts. Consequently there is no regular cattle rotation. It is used for young cattle, sheep and robust horses. The intensive grazing comprises the strip grazing and the intensive set stocking. On an intensive set stocking pasture the grass is kept short (5 - 7 cm) through permanent grazing without cattle rotation. Strip grazing provides a flexible paddock size by a mobile electric fence. A temporary fence line is moved across the area. Depending on the conditions a back fence line is moved in some cases to prevent cattle from entering previously grazed strips. The rotational grazing, which is also called paddock grazing, can be performed extensively or intensively depending on the number of paddocks and grazing days per cattle drive. [26]

B. Seiferth, *Development of a system for selective pasture care by an autonomous mobile machine*, Fortschritte Naturstofftechnik, https://doi.org/10.1007/978-3-662-61655-0_5

5.2   General process environment and sequence

As described in 5.1, there are grazing systems with cattle rotation. Due to the fact that automated operations involve risks of collisions with living beings these systems is suited in the first instance. The risks of conflicts between cattle and the machine are not given. Thus, the rotational grazing and strip grazing are relevant systems, as highlighted in Figure 16.

With strip grazing, the fence is moved daily or after up to four days depending on the grass growth. A strip area of 0.004 up to 0.01 hectare are needed for one living unit (LU), which corresponds to the weight of a mature cow of 500 kg in husbandry. With the rotational grazing the pasture area is subdivided in 5 to 20 paddocks. The boundaries of the single paddocks usually do not change during the grazing season. The grazing period per paddock lasts from 2 to 7 days. The stocking rate $s_{Paddock}$ amounts 0.02 to 0.09 hectare per LU. [26] [66]

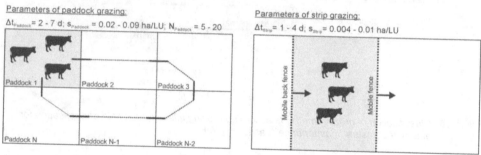

Figure 17: Paddock grazing (left); strip grazing (right) (data according to [26], [66])

Finally, these guide values for the stocking range, grazing period per strip or paddock and the number of paddocks per herd define the grazing performance of the cattle. Thus, they are important with regard to the dimensioning of the maintenance machine. Its maximal area performance must be higher than the maximal area performance of the grazing cattle herd. After animals having grazed and left an area, the pasture maintenance machine has to become active.

The automated pasture maintenance can be performed in different ways. One concept is to separate the scouting process and the working process. That means the whole area is scouted to localise relevant areas, which have to be mulched or reseeded, in the first step. In a second step, a machine goes to the particular areas and mulches or reseeds there. For this, the optimal route can be calculated so that the covered distance can be as short as possible. The two tasks scouting and working can be performed by the same machine or by two different task-optimised machines. In the latter case, the vehicle for scouting is lightweight and equipped with adequate sensors for data acquisition, the other one for working is equipped with the mulcher and seeder. The use of an UAV for scouting is also conceivable.

Another concept is to do scouting and maintenance operations (mulching and reseeding) at once in one step. The machine scouts the paddock and directly mulches or reseeds the localised un-grazed or soil area. The machine is equipped with both adequate sensors for data acquisition and implements.

This work deals with the last-mentioned concept with one vehicle, because the focus can be on the tasks itself in the first instance. The other concept with two vehicles can be deduced from this solution. The solution with UAV is out of scope in this work.

Thus, the machine has to scout the whole grazed area for leftovers as well as damages by footsteps in the sward and mulch or reseed them immediately at the same time. This process has to be finished in maximal two days because of the plant growth. In this phase spots, which have not been grazed by the cattle, must be mulched afterwards (see paragraph 1.3). Figure 18 shows the three phases grazing, maintenance and growth period of one paddock.

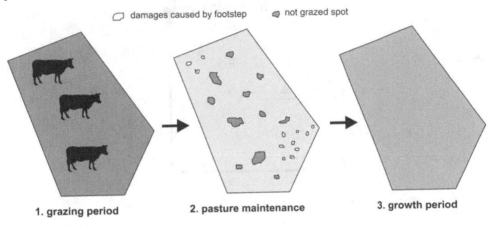

Figure 18: Sequence of one paddock

According to the information provided by [67] the averaged herd size among commercial dairy producers comprised 116 cows in countries with an averaged herd size of more than 20 animals. Resultant the defined maximal herd size the machine is designed amounts 100 LU. Finally, the minimum area performance of cattle on pasture $\dot{A}_{cattle}$ can be calculated as follows:

$$\dot{A}_{cattle} = \frac{S_{Paddock} \cdot 100\ LU}{\Delta t_{paddock}} = \frac{0.09\ \frac{ha}{LU} \cdot 100\ LU}{48\ h} = 0.19\ ha/h \qquad (5.2.1)$$

5.3   Mulching, seeding and transit process

After having calculated the maximum area performance of the cattle or rather the necessary area performance of the machine for automated pasture maintenance in paragraph 5.2 by taking into account the given grazing process, now the required operation parameters can be defined. The following model simulates exemplary a grazing farm with 100 LU, a maximal theoretical necessary paddock area of $A_P = 0.09$ ha and a number of $N_P = 20$ paddocks (see paragraph 5.2). This scenario is used to determine the scale of operating parameters, which guarantee a sufficient performance of the pasture maintenance machine in worst case. The layout of the squared paddocks is pictured in Figure 19. These paddocks are grazed six times (N_C) per season which usually lasts at least $N_d = 150$ days (in Europe e.g. from May to September) [26]. After a certain period of time Δt_O (operational time) the machine has to go back to the service station for refueling and servicing for a period of Δt_S (service time).

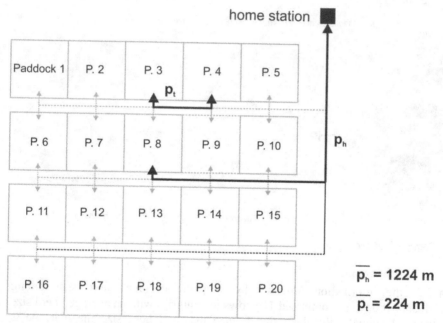

Figure 19: Simulation model of a grazing farm

The required paddock area performance of the machine $\dot{A}_{machine}$ for this model can be calculated as follows considering the whole grazing season:

$$\dot{A}_{machine} = \frac{\text{total area per season}}{\text{total time per season} - \text{necessary time for refueling or charging and transfers}} \quad (5.3.1)$$

Thereby the path for transit between two paddocks p_t is corresponding to the side length of one paddock with a square shape of 224 m and the path lengths for coming home from each paddock to the docking station p_h are averaged:

$$\bar{p}_t = \frac{\sum p_{ti}}{N_P} \qquad (5.3.2) \qquad\qquad \bar{p}_h = \frac{\sum p_{hi}}{N_P} \qquad (5.3.3)$$

The operation time of the machine Δt_O is limited by the energy storage. Besides the parameter Δt_O, as well as the time for service Δt_S, the time for recharging or refueling and the speed for transit v_t have influence on the necessary paddock area performance. The following formula (5.3.4) expresses the area performance with detailed grazing parameters:

$$\dot{A}_{machine} = \frac{N_P \cdot N_C \cdot A_P}{N_d \cdot t_{day} - \left[N_d \cdot \left(\frac{t_{day} - \Delta t_S}{\Delta t_O} \cdot \frac{p_h}{v_t} \right) + N_d \cdot t_s + (N_C \cdot N_P) \cdot \frac{p_t}{v_t} \right]} \qquad (5.3.4)$$

This expression results in the following relationship between transport speed and operational speed:

$$v_t(v_o) = \frac{N_P \cdot N_C \cdot \bar{p}_t + N_d \cdot \left(\frac{t_{day} - \Delta t_S}{\Delta t_O} \cdot \bar{p}_h \right)}{\frac{N_P \cdot N_C \cdot A_P}{v_o \cdot w} - N_d \cdot t_{day} - N_d \cdot \Delta t_s} \qquad (5.3.5)$$

The following diagram shows this relationship for three different working widths and the influence of the operational time Δt_o based on the state of the art of remote controlled mowers (see paragraph 2.2).

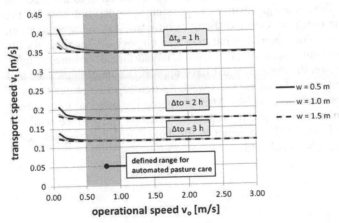

Figure 20: Relationship between required transport speed and operational speed and the influence of the operational time

The necessary transport speed does not change significantly with the operational speed. Generally the required transport speeds v_t is quite low compared with usual agricultural grassland machinery. The operational speed and thus, the area performance can be also quite low. In regard to functional safety and operational costs this circumstance promotes the possibility of automated maintenance operations on pastures. Risk issues in regard to operational safety can be managed better with low operational speeds. Finally, the defined operational speed for automated pasture care is in a range between 0.5 to 1 m/s.

5.3.1 Detection of mulching and seeding spots

In theory, leftovers can consist of one leaf of grass or can be larger. In the first instance it is not necessary to detect every single left blade of grass. Damages of the sward because of footsteps have to be detected. Footsteps of cattle have an area of approximate 100 x 100 mm. Consequently the defined minimal size of spots, where maintenance is required and which are to identify, is 100 x 100 mm. Furthermore on pasture a maximal grass height of leftovers of 200 mm is expected. The defined minimum grass height of leftovers is 50 mm.

5.3.2 Mulching process

In this section, target requirements to the process for removing un-grazed spots after grazing on pasture are defined to evaluate the different systems (see paragraph 2.1) and finally to select one of them. Generally, a low power requirement of the mulcher must be aimed. From the energetic point of view and in regard to soil compaction low mass of the mulcher is to prefer. Another very important aspect for the operation of the machine is high operational reliability. Foreign matters, e.g. stones, must not restrict the functionality or damage the machine. Under difficult conditions, e.g. lying vegetation, the machine must not block with material, what can lead to loss of function. As the detection of soil profile on pasture (section 5.4.2) shows, the surface of pastures is quite rough and hilly. For this reason the machine must have appropriate height guidance and has to be shortly designed to avoid damaging the sward. As described large quantities of grass have to be chopped as finely as possible. After chopping the material must spread widely [50]. Accumulation and clumping of vegetable material have to be avoided, especially under wet conditions. Finally, damage to the fauna (insects, mousses, toads etc.) by the mulcher tools must be avoided [51].

5.3.3 Seeding process

Special mixtures of seeds are applied for reseeding on pastures. The seed mixtures usually consist of grass, legumes and herbs. For a successful seeding the seeds have to be sown on the surface, maximal at a soil depth of 5 mm. The flatter the seed placement into the soil, the shorter is the germination time. [68]

Broadcast seeding will respond better to the task of automatic pasture maintenance, because it is not susceptible in regard to abrasion and material blockages in comparison to drill seeding. The seed rate s of broadcast seeding is 5 to 10 kg/ha [26]. The seed container must have a capacity which guarantees seeding during the whole operation of the machine on pasture until refueling or recharging is necessary. Furthermore the seed rate must be adjustable variably or incrementally for the farmer due to the different necessary seed rates.

5.4   Mobility of the automated pasture care machine

5.4.1 Vehicle control and navigation

On pastures there is no structure like in plantations which can be used for navigation. For this reason the vehicle has to follow predefined trajectories which fully or partially cover paddock areas, e.g. by parallel paths, where maintenance operations have to be carried out. A precision of few centimeters (depends on working width of implements) is necessary considering the lateral deviation. The user has to have access to the machine at any time to control manually or at least monitor it from a stationary supervisor PC. Consequently a wireless connection must always exist.

5.4.2 Vehicle chassis

The terrain can be hilly and its surface can be rough because of treading cattle. The properties of terrain are important for dimensioning the machine for automated pasture maintenance. The mobility of the machine must be guaranteed at any time. Thus, the soil profile was detected to define requirements. For this, measurements have been performed on exemplary pasture areas in Upper Bavaria. The grazing farm was in the alpine upland with hilly pasture areas. A carriage which was pulled by a small tractor was used (see Figure 21). In principle the soil profile was sensed by the wheels of one axle rolling over the pasture area at the lowest possible speed. This method is sufficient to detect the profile of surface although edges on the surface cannot be detected because of the diameter of the wheels. The roll and pitch angles Θ_x and Θ_y of this measuring axle as well as its GPS positions (GPS-RTK) have been measured. This raw data has been detected by a laptop and the mathematical analysis has been performed by post-processing. Due to the height of the GPS antenna Δz_{v_a} the first step was to calculate the GPS positions of the central point C on the connecting line between the soil contact points A and B of the two wheels. The sample rate of the GPS measurement was 10 Hz and the measuring frequency of the inclination detection was 40 Hz, so that the resulting resolution is between 0.7 until 0.8 points/cm according the travelled distance by applying an interpolation.

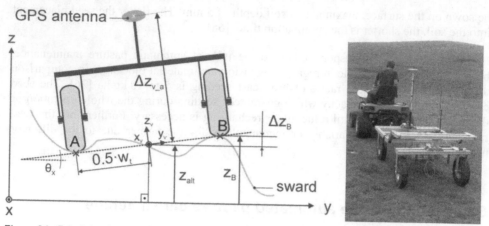

Figure 21: Principle of soil profile detection (right); Detection of soil profile on pastures by a pulled small carriage (left)

Then the positions of one soil contact point A or B are calculated by the corresponding role angles θ_{xi}:

$$z_{Bi} = z_{alt} + \Delta z_{Bi} = z_{alt} + \sin(\theta_{xi}) \cdot \left(\frac{w_t}{2}\right) \tag{5.4.2.1}$$

Finally, these calculated individual points result in the soil profile of the measured pasture areas. The detected soil profile serves to define the parameters of the chassis, so that a stable motion on pasture while not damaging the sward is ensured. Consequently a skid steered vehicle and a solution with tracks are excluded. A mobile machine with a wheel chassis is considered. First the necessary ground clearance of the vehicle is determined. It depends on the wheelbase and the track gauge of the vehicle. Figure 22 shows the principle of the calculation.

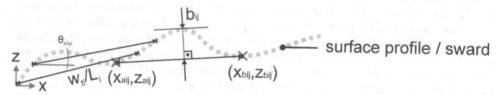

Figure 22: Principle of determination of the necessary ground clearance

All possible pairs of points for different track gauges w_{ti} and wheelbases L_i are analysed and each necessary ground clearance b_i is calculated on the basis of the equations in Annexe 1. The results are shown in Figure 23, where all maximum ground clearance b_{imax} are figured. As expected the necessary ground clearances are generally higher for large track gauges or wheelbases.

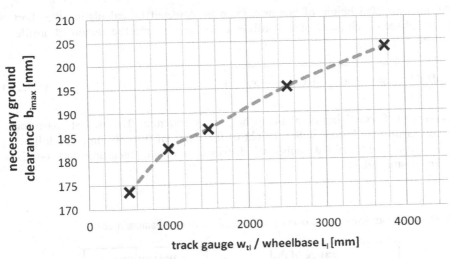

Figure 23: Necessary ground clearance depending on track gauge and wheelbase

Another feature of off-road vehicles are pivoting axles which are characterised by the ramp travel index. A permanent contact between all wheels and the soil is important in regard to the measurements by the 2D laser scanner. A pivoting axle adapts to the ground and specially compensates unevenness or bumps on the ground. For calculating the dimension of bumps on pasture and thus, the necessary minimal ramp travel index the detected soil profiles have been smoothed by a filter applying the half width moving average method of twelve values, which corresponds to a travelled distance of a half meter, to get the profile of terrain without the noise caused by surface unevenness. The following Figure 24 shows an example of the two calculated profiles.

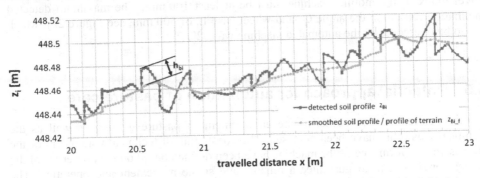

Figure 24: Example of a detected soil profile consisting of detected measurement values (dark grey line) and of filtered measurement values (light grey line)

Finally, an estimated height of bumps h_{bi} can be detected by calculating the shortest distances between each point of the unfiltered soil profile and the smoothed profile of terrain (see Annexe 1).

$$h_{bi} = MIN \left[\sqrt{(z_{Bi\_f} - z_{Bi})^2 + (x_{Bi\_f} - x_{Bi})^2} \right] \qquad (5.4.2.2)$$

The maximum detected height of bump h_{bmax} was 190 mm. The largest share (circa 97 %) concerning the bumps with a height of the range of 0 to 50 mm. The following Table 2 shows the resulting individual distribution of the measurements on these exemplary pasture areas.

Table 2: Resulting individual shares of bump heights on exemplary pasture areas

range of $|h_b|$	percentage
0 mm < h_b < 50 mm	97.37 %
50 mm < h_b < 100 mm	2,6%
100 mm < h_b < 150 mm	0.08 %
h_b > 150 mm	0.03 %

Finally, the dimension of the necessary ramp travel index rti is twice the maximum detected height of bump in the ground which the chassis has to adapt in worst case.

$$rti_{min} = 2 \cdot h_{bmax} \qquad (5.4.2.3)$$

Due to the very small share of height of bumps higher than 50 mm the maximum ramp travel index of the mobile machine must be at least 100 mm. The maximum detected inclination (based on the angle Θ_y and wheel base L_i of 500 mm, see Figure 22), which has to be overcome by the automated machine was 32°.

5.5   Potential approach for automation of pasture care

In this chapter the constraints of the environment of pasture grazing as well as the technical feasibility have been analysed. The constraints of technical parameters of the chassis and performance of the machine have been defined based on the properties of the process environment to guarantee a reliable and stable movement and operation. The technical relations of the detection by a 2D laser scanner and possible challenges have been recognised. The result is the following approach:

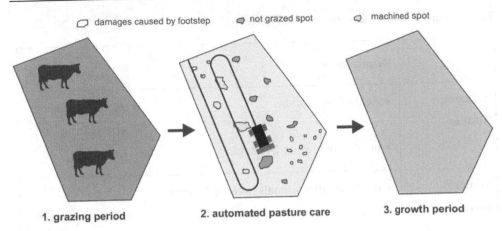

Figure 25: Process of automated pasture care for one paddock consisting of three phases

The process of rotational grazing provides a time window for automated mechanically operations on the paddock (see paragraph 5.2). The machine can perform its operation during the phase between cattle leaving the area and growth period. This phase lasts maximal two days (5.2).

Table 3: requirements for automation of pasture care

	Requirement	
Grazing system	• field of application:	rotational grazing strip grazing
	• number of grazing livestock:	100 LU
General process	• operation after animals leaving the area	
	• pasture maintenance within	two days/paddock
	• minimum transportation speed:	0.5 to 1.0 m/s
	• minimum area performance:	> 0.75 m²/s
Spot detection	• minimum grid resolution:	100 x 100 mm
	• minimum grass height of leftovers:	50 mm
	• maximum grass height of leftovers:	200 mm
	• scanning width	= working width
Mulching	• maximum chop length:	50 mm
	• chopped material must be spread widely	
Seeding	• range of seed rate	5 to 10 kg/ha
Vehicle control and navigation	• guidance	predefined trajectories
	• precision of navigation	5 cm
Chassis	• ground clearance	> 175 mm
	• ramp travel index	> 100 mm

6 Evaluation, selection and development of components

6.1 Mobile platform

The machine is developed based on conventional technology, in particular based on a commercially available machine. The research on the market was focused on remote controlled machines without a driver's seat or cabin equipped with a mulcher or to which a mulcher and additional implements can be attached (see paragraph 2.2). These machines are initially intended for machining public green areas. The working widths usually are between of 0.5 to 1.5 m and the operational speed corresponds to walking speed to guarantee walking behind the machine for control. Consequently the area performance is between maximal 0.45 ha/h and 1.5 ha/h if an operational speed between 3 km/h and 10 km/h is assumed. This range of area performance is above the required area performance for automated pasture maintenance (see paragraph 5.5). Finally, several machines on the market have been analysed if they fulfill the required constraints. Possible vehicles have been compared according cost-effectiveness and finally the machine "RoboZero" of the manufacturer Energreen was selected as base machine for automated pasture care (see Figure 26).

Technical specifications	
Engine:	YANMAR 3-cylinder diesel engine 24 kW power 1.331 cubic capacity
Drive:	hydraulic drive
Speed:	up to 11 km/h
Dimensions:	450 / 580 / 1090 mm (length / width / height)
Weight:	780 kg (plus implements)

Figure 26: Base machine "RoboZero" for automated pasture care [29]

The vehicle is equipped with a pivoting front axle with a ramp travel index of $rti = 150$ mm. The ground clearance of the chassis with a wheelbase of 1.35 m is $b = 150$ mm. This is sufficient for the prototype, although a ground clearance of at least 180 mm is required for this wheel base (see paragraph 5.4.2). The machine is a front-steering vehicle. The flail mulcher can be attached between the front and rear axle of the vehicle. The rotor drum is equipped with Y-flails.

© The Editor(s) (if applicable) and The Author(s), under exclusive license to Springer-Verlag GmbH, DE, part of Springer Nature 2020
B. Seiferth, *Development of a system for selective pasture care by an autonomous mobile machine*, Fortschritte Naturstofftechnik,
https://doi.org/10.1007/978-3-662-61655-0_6

The operating time Δt_O is necessary to determine the range of the operational speed on the paddock area as explained in paragraph 5.3. The operational time Δt_O can be estimated by manufacturer information about the specific fuel consumption of the engine b_e and the maximal fuel tank capacity V_f from the manufacturer (see [29]):

$$\Delta t_o = \frac{V_f}{b_e} \qquad\qquad\qquad (6.1.1)$$

As calculated in Annexe 2, the operational time Δt_O is approximate 3.5 h. Thus, the required area performance according to equation (5.3.4) is $p_p = 0.75$ m²/s. The operational required speed v_o is 0.58 m/s corresponding to equation (6.1.2).

$$v_o = \frac{p_a}{w} \qquad\qquad\qquad (6.1.2)$$

The machine is driven hydraulically. Figure 27 shows a scheme of the hydraulic drive system. The main pump block is flanged at the Diesel engine. In principle there are three circuits. One circuit is for the drive of the rear wheels. It is supplied by a double piston pump. Each rear wheel is driven by a separate hydraulic motor. The differential balance can be blocked by valves. The other circuit is used for the mulcher drive. Another circuit serves for the control of the hydraulic cylinders. One is a single acting cylinder which is used to elevate the mulcher. The mulcher moves down realized by gravity force. The second double acting cylinder is for steering the front wheels. A hydraulic line supplies the rear wheel brakes. Under pressure the brakes are released. Without pressure they are active.

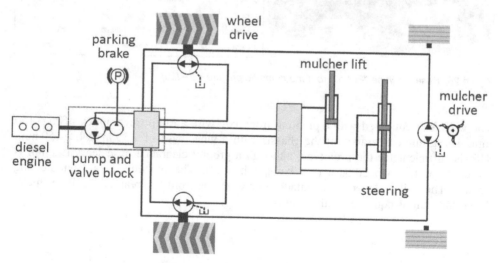

Figure 27: Hydraulic scheme of the machine

The machine is originally radio remote controlled using ultra high frequency (UHF). On the machine is a receiver unit which converts signals into CANopen commands which are used to communicate with the main machine control unit (see Figure 28). First the CAN messages of Process Data Objects (PDO) with own Communication Object Identification (COB-ID) consist of eight bytes. For example, the byte number 3 of the COB-ID 0x000 is reserved for the steering function. If its value $cmd_{steering}$ is 127 the wheels are in neutral position, which means the vehicle moves on a straight line. A value of 0 means steering left and 255 steering right. Thus, there are 127 digits for steering in one direction.

Figure 28: Layout of the original machine control

In order to measure the fuel consumption (see paragraph 9.3), an extra Diesel tank was installed and integrated into the fuel system which can be weighed before and after a test run (see Figure 29). The volume of this tank was 5 l, so that time consuming measurements were not necessary. Two ball valves enabled a Diesel supply from the original tank via the new tank or only from the installed extra tank. This tank was weighted by a hanging scale.

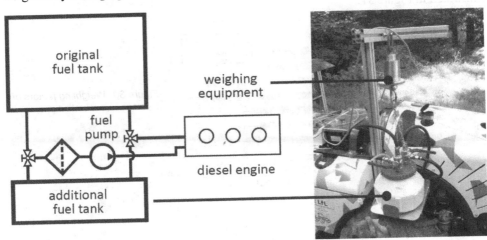

Figure 29: Diesel circuit system (left); installed weighing device (right)

6.2 Vehicle navigation

As mentioned in section 4.1, the GPS has enhanced in localisation of agricultural machinery on free field operations. Thus, a GPS-RTK or DGPS system is well suited for automated maintenance operations on pasture areas. The accuracy of those systems will be sufficient.

6.3 Implement for processing un-grazed spots on pasture

As described in section 2.1, there are different types of implements for removing leftovers on pasture. Possible types of mulcher or rather mower are in principle machines with rotational or oscillating tools. Machines with rotational tools include the rotary mulcher and the flail mulcher. The double cutting bar has two horizontal transverse to the direction of travel positioned oscillating knifes. The following Figure 8 shows the described types of machines.

In chapter 5.3.2 the requirements to the process for removing un-grazed spots on pasture are defined. These requirements are the criteria for the evaluation of the three types of machines. The evaluation criteria have been weighted by pairwise comparison. Figure 30 4shows the results of weighting the evaluation criteria. The evaluation itself was done in tabular form based on statements of literature (see Annexe 3). Ten experts have been interviewed for pairwise comparison according the weighting of evaluation criteria and have given their scoring points for this evaluation. The result of the evaluation is illustrated in Figure 31.

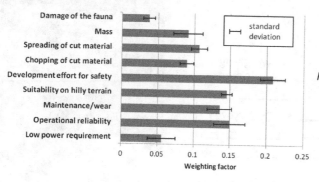

Figure 30: Weighting factors of evaluation criteria

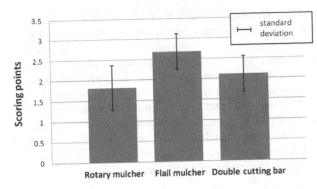

Figure 31: Result of the evaluation of mowing devices

The flail mulcher has gotten the most scoring points in the evaluation. Thus, it is the most suitable one for mulching operations on cattle pasture.

6.4 Implement for reseeding spots on pasture

As described in section 5.3.3, broadcast seeding is to be used for the automated pasture maintenance. Possible seeder types for broadcast seeding are spreaders with a rotating distributing plate or spreaders with a horizontal spiral conveyor shaft in a box. Another option is a usual drill seeder without the tools for tillage (see paragraph 2.1). At this point a complex evaluation is waived because the expected important positive effects of selective broadcast seeding instead of seeding the whole area are limited to saving seeds. Expected energy savings because of selective seeding operations will not be significant considering the complete system including mulching operation. Due to the circumstance, that the minimal spreading width of conventional commercial available spreaders with distributing plates is 2 m and the spreading width must correspond to the mulching width (see paragraph 5.3.3), the decision is taken to design a seeder based on the ILT fertiliser spreader for plots [69]. The grains are applied by cellular wheels and fall through downpipes to the ground.

Figure 32: ILT fertiliser spreader for trial plots [60]

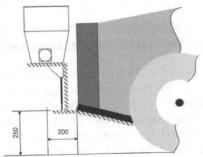

Figure 33: Design of the seeder for automatic seeding operations on pasture

The same principle and concept of this fertiliser spreader for trial spots was adopted and modified for the application as seed spreader for reseeding spots on pastures (see Figure 33). Preliminary tests had shown that seed mixtures consisting of different grains with different sizes and masses do not separate by vibrations and impacts while travelling on rough terrain surface. Thus, the use of a mixing device was waived. The distance between the downpipe outlets was reduced to 90 mm. Thus, a working width of 1300 mm (see paragraph 6.1) results in 15 downpipes and seeding units. The total width of the seeder must not exceed the total width of the machine of 1400 mm. As explained in 5.3.3, the maximal operating time Δt_O of the machine, until refueling is necessary, is needed for dimensioning the seed container. Finally, the maximum volume of the seed container can be determined considering the operating time Δt_O:

$$V_s = \frac{\dot{A}_{machine} \cdot \Delta t_o \cdot s}{\rho_s} = \frac{0.27 \, \frac{ha}{h} \cdot 3.5 \, h \cdot 10 \, \frac{kg}{ha}}{0.465 \, \frac{kg}{l}} = 20 \, l \qquad (6.4.1)$$

The shaft with the cellular wheels is electrical driven by a step motor. The seeder was mechanically mounted at the vehicle rear by a mounting frame, which was fixed to the main machine frame. Thereby the ground clearance of the vehicle had to be guaranteed, especially for operations in hilly terrain (see Figure 33).

6.5   Sensor solution for spot detection

Considering the comparison in Table 1, modern laser scanning technology has a good basis in regard to the task of spot detection for pasture care automation (see paragraph 3). In fact, several scientists focus on topics concerning laser scanner technology in agriculture for navigation or plant phenotyping ([70], [70], [71]). The laser scanning technology is to use to identify spots requiring pasture maintenance, in particular to mulch leftovers and to reseed damages caused by footprints left after grazing. Thus, selective operations can be performed immediately after being detected.

6.5.1 Definition of installation and operational parameters

The use of a 2D laser scanner sets framework conditions to the technical system. The position of the scanner in relation to the vehicle, especially the attaching height, inclination, location and the width of the scanning zone, angular resolution and the scanning frequency have influence in defining operation parameters of the machine. In this section first theoretical relationships between operation parameters of the 2D laser scanner and the mobile platform are shown. A detection of maintenance spots on pasture by a 2D laser scanner is only possible if it is attached at the front of the machine in a manner that the scanning plane intersects the ground. Figure 34 shows generally a possible installation of a 2D laser scanner at the front axle in centre position.

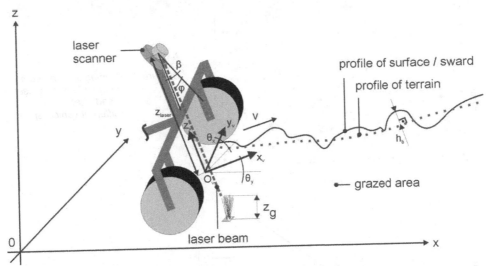

Figure 34: General installation of a 2D laser scanner at the front of a vehicle

Generally, the measuring points result in parallel curves on the ground level because the overlay of both linear motion of the vehicle and the rotational movement of the laser head with a frequency f. These curves can be described mathematically as follows on the assumption that the ground level x-y plane is flat (see Annexe 4):

$$\begin{bmatrix} x_j \\ y_j \end{bmatrix} = \begin{bmatrix} x_0 + j \cdot v_x \cdot \frac{f^{-1}}{360°} \cdot \Delta\varphi \\ \frac{z_{laser}}{\cos(\beta)} \cdot tan\left[arctan\left(\frac{0.5 \cdot w_{scan} \cdot \cos(\beta)}{z_{laser}}\right) - j \cdot \Delta\varphi \right] \cdot z_{laser} \end{bmatrix} \qquad (6.5.1.1)$$

$$\text{with } j \in \left[0; 1; 2; \dots ; \frac{2 \cdot arctan\left(\frac{0.5 \cdot w_{scan}}{z_{laser}}\right)}{\Delta\varphi} \right] \qquad (6.5.1.2)$$

If the ratio between the driving speed v_x and the scanning frequency (rotational speed of the laser head) f is low, these curves approximate to straight lines. Figure 35 shows different curves of measuring points for different driving speeds. The distance between the curves Δx_{res}, which characterises the resolution in direction of travel x, depends on the scanning frequency and the driving speed:

$$\Delta x_{res} = v_x * f^{-1} \qquad (6.5.1.3)$$

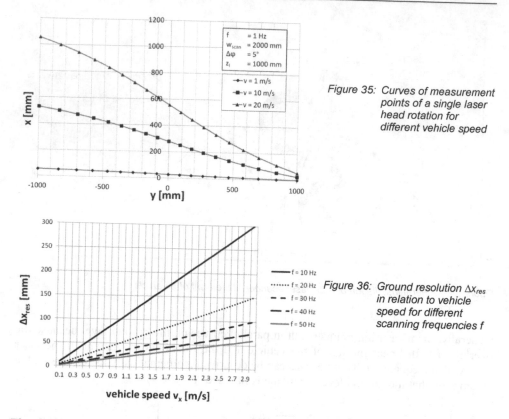

Figure 35: Curves of measurement points of a single laser head rotation for different vehicle speed

Figure 36: Ground resolution Δx_{res} in relation to vehicle speed for different scanning frequencies f

The diagram in Figure 36 contains dependency between Δx_{res} and the vehicle speed for different scanning frequencies f. Besides the resolution in direction of travel Δx_{res} the resolution in transverse direction Δy_{res} must be considered, too. It corresponds to the distance between the successive rotating reflected laser beams on the ground. The maximum distance between two measuring points Δy_{max} on ground level is between both external laser beams of the scanning width. The position of the scanner in relation to the vehicle, especially the attaching height z_l, the inclination β, the width of scanning zone w_{scan} as well as the angular resolution $\Delta\phi$ have influence on Δy_{max} and finally on the resolution Δy_{res} (see formula (6.5.1.4) and Annexe 4).

$$\Delta y_{max} = \frac{w_{scan}}{2} - \frac{z_{laser}}{\cos\beta} \cdot \tan\left[\arctan\left(\frac{0{,}5 \cdot w_{scan} \cdot \cos\beta}{z_{laser}}\right) - \Delta\varphi\right] \tag{6.5.1.4}$$

The following diagram in Figure 37 shows this distance depending on the parameters w_{scan} and z_{laser}.

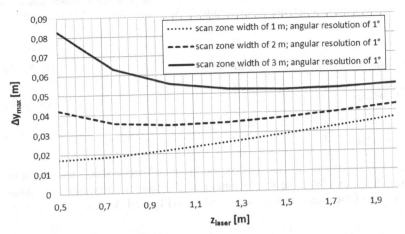

Figure 37: Resolution in y-direction depending on the width of scanning plane w_{scan} and the installation height of the 2D laser scanner z_{laser}

The larger the width of scanning zone w_{scan} the larger is the distance Δy_{max}, the lower the size of the resolution Δy_{res} on ground level. A higher installation height z_{laser} of the laser scanner must not necessarily result in a higher distance Δy_{max}. If higher objects on the ground, e.g. un-grazed spots on pasture, are scanned, a so-called laser shadow occurs. This shadow describes an area, which cannot be scanned by the laser scanner and relevant spots can possibly not be detected (see Figure 38).

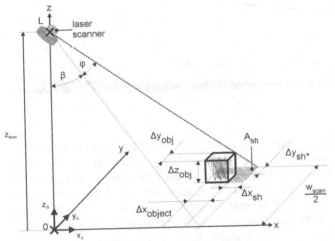

Figure 38: Occurring laser shadow during scanning higher objects on the ground

Both the installation height z_{laser} of the laser head, the inclination of the scanning plane β, the width and the height of the object Δy_{obj} and Δz_{obj} as well as the width of the scanning zone w_{scan} have particularly influence on the dimensions of the maximum laser shadow. They include the dimensions of the laser shadow Δx_{sh} and Δy_{sh*}. All these three parameters characterise the area of the shadow A_{sh}.

$$A_{sh} = \Delta x_{sh} \cdot (\Delta y_{obj} + \Delta y_{sh*}) + \Delta x_{obj} \cdot \Delta y_{sh*} \qquad (6.5.1.5)$$

$$\Delta x_{sh} = tan(\beta) \cdot \Delta z_{obj} \qquad (6.5.1.6)$$

$$\Delta y_{sh*} = \Delta z_{obj} \cdot \left(\frac{0.5 \cdot w_{scan}}{z_{laser}} \right) \qquad (6.5.1.7)$$

The length Δx_{sh} of the laser shadow depends on the laser inclination β and the height of the object Δz_{obj} (see Figure 39). The displacement of the shadow in y direction Δy_{sh*} depends on the width of the scanning zone w_{scan} and z_{laser} (Figure 40).

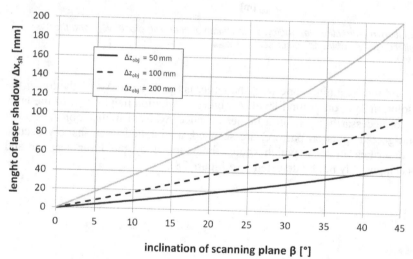

Figure 39: Length of laser shadow Δx_{sh} depending on the laser inclination β and height of scanned object Δz_{obj}

It is maximal within the scanning zone if the outer shadow border is on the edge of the scanning width. For that reason this case is considered, like it is illustrated in Figure 38. Figure 40 shows the influence of the installation height of the laser head z_{laser} on the lenght of the laser shadow Δy_{sh*}. There is no significant change of the laser shadow width Δy_{sh*} from a height of $z_{laser} > 1000$ mm for widths w_{scan} lower than 2000 mm. Thus, an installation height of the laser scanner z_{laser} in this range is to target. Smaller scanning widths w_{scan} result in smaller laser shadow displacements Δy_{sh*}.

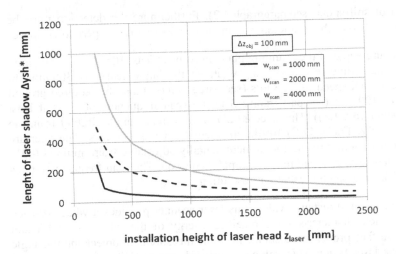

Figure 40: Laser shadow displacement $\Delta y_{sh}{}^*$ depending on the laser inclination β and height and width of scanned object Δz_{obj} and Δy_{obj}

Since the relations of different parameters of scanning the ground have been analysed, requirements for localising spots on pasture, where maintenance operations are necessary, can be determined. Thereby the resolution on the ground surface is the important criterion. In theory leftovers can consist of one leaf of grass or can be larger. In the first instance, it is not necessary to detect every single left leaf of grass. The defined maximum resolution is orientated towards damages of the sward because of footsteps which have to be detected. Footsteps of cattle have an area of approximate 100 x 100 mm. Consequently the defined minimal size of spots, where maintenance is required and which are to identify, is 100 x 100 mm. Thus, the resolutions on ground in both the direction of travel Δx_{res} and the lateral direction must be at least 100 mm. Furthermore on pasture a maximal grass height of leftover of 200 mm is expected. Based on these constraints technical parameters of the 2D laser scanner can be defined. The attaching height of the 2D laser scanner at the front of the vehicle will be more than 300 mm. In this context the aperture angle must allow the scan of a ground area with a width according to the working width of the mulcher. In regard to the operational speed and the required ground resolution x_{res} of 100 mm the scan frequency must be at least the frequency according formula (6.5.1.8).

$$f = \frac{2 \cdot v_x}{\Delta x_{res}} \qquad\qquad (6.5.1.8)$$

Tests have to show which angular resolution is necessary. For this reason, a laser scanner with a high angular resolution must be used for the prototype to find out the optimal resolution $\Delta\varphi$. In this context, the light spot diameter of the laser beam depending on the distance between emitter and ground surface must also be considered. Finally, the scanner must provide another value, besides the output distance, which characterises the received energy of the reflected laser beam depending on the surface properties to use it for the detection of soil spots (see paragraph 4.2). Furthermore the detection must be performed in real-time to enable immediate maintenance operations at operational speed.

The maximal aperture angle of the scanning plane must be at least 105° (z_{laser} = 500 mm, w_{scan} = 1300 mm) according to the working width of the implements of 1300 mm. In regard to the operational speed of 0.58 m/s (see paragraph 6.1) and the required ground resolution Δx_{res} of 100 mm (see paragraph 5.3.1) the scan frequency must be at least 11.6 Hz (see equation (6.5.1.8)). The selected 2D laser scanner (model R2000 from the manufacturer Pepperl & Fuchs [72]) fulfills all mentioned requirements. The aperture angle is 360° and the angular resolution is incrementally adjustable from minimal 0.01° to maximal 5°. Thus, tests can lead to the optimal resolution. The scan frequency ranges from 10 to 50 Hz depending on the angular resolution.

In addition to the measured distance value, this laser scanner provides a value for each measuring step, which characterises the received energy of the reflected laser beam depending on the surface properties, distance from the measurement object and the angle of incidence of the laser beam. This value of received energy (VRE) has no unit and ranges from 0 to 4095 digits. [72]

This value is used to distinguish between ground covered with grass and soil in order to detect spots on pasture where maintenance operations are required.

Technical specifications	
Light emitter:	laser diode
Illuminant:	red, alternating light, 660 nm wave lenght
Scan rate:	10 .... 50 Hz
Measurement mehtod:	pulse ranging technology (PRT)
Dimensions:	118 / 106 / 117 mm (length / width / height)

Figure 41: 2D laser scanner R2000 from Pepperl & Fuchs [72]

As explained, both the height z_{laser} and the inclination of the scanning plane β have influence on the dimensions of the laser shadow. The following diagrams show the permissible attaching heights z_{laser} and inclinations β considering the defined minimum size of spots of 100 x 100 mm which have to be detected (see paragraph 5.5). The minimum installation height of the laser head z_{laser}, calculated based on equation (6.5.1.4), is 1250 mm (see Figure 42). The maximum inclination angle β is 27° (calculated by equation (6.5.1.6)) (see Figure 43).

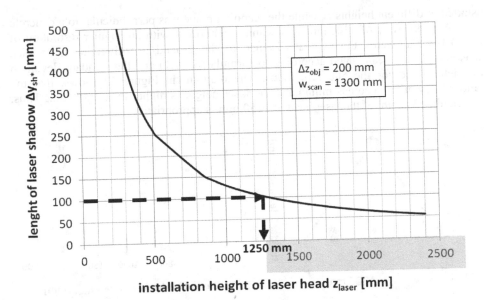

Figure 42: Permissible range for the installation height of the laser head z_{laser}

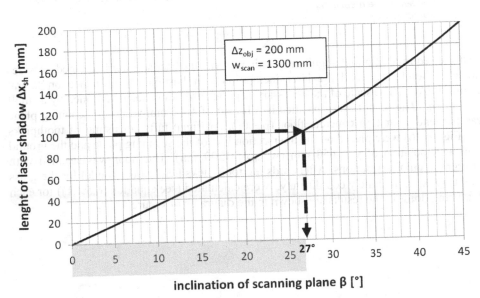

Figure 43: Permissible range for inclinations of the scanning plane β

Due to the fact that the precision of output measured distance is affected among other things by the distance to the reflecting object [72], a test validates this circumstance and serve to determine an optimal installation height of the laser scanner. The scanner was installed at different heights z_{li} while the scanning plane was perpendicular to a concrete floor. A width w_{scan} of 1300 mm was scanned 500 times with an angular resolution of $\Delta\phi = 0.1°$. Afterwards the vertical distance to the ground in z-direction h of each measured value is calculated according to formula (6.5.3.1) and the standard deviation was determined for the 500 repetitions. The diagram in Figure 45 shows, that the standard deviation becomes minimal if the laser height is higher than 1000 mm. Thus, the calculated height of the laser head of 1250 mm is appropriate.

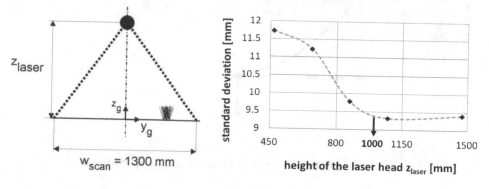

Figure 44: Tests draft of analysis of the influence of laser height zlaser on the measuring precision

Figure 45: Result analysis of the influence of laser height zlaser on the measuring precision

Next the required minimal angular resolution of the 2D laser scanner is defined. Besides the laser height z_{laser} and the laser inclination β, it has mainly influence on lateral resolution Δy_{res}. The scale of the necessary angular resolution was determined by a small experiment. The scanner was installed in a height of 1000 mm while the scanning plane was perpendicular to a concrete floor (β = 0°). Artificial tufts consisting of toothpicks (see Figure 46) were placed on the intersection line of the scanning plane and the ground level at a position of y = 650 mm (w_{scan} of 1300 mm). The floor was scanned with different angular resolutions. Afterwards the toothpicks height (as a model for the grass height) grass z_{gi} and the y coordinate y_{gi} were calculated (see equation (6.5.1.1)) for each measuring point i. The results are shown in the diagrams of Figure 47.

Figure 46: Artificial grass tufts consisting of toothpicks

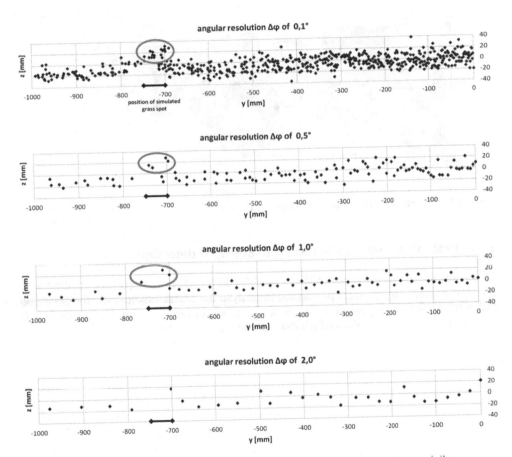

Figure 47: Results of the experiment to determine the scale of necessary angular resolution

The measuring points related to the artificial tuft are grey circled. The artificial tuft is clearly visible if the angular resolution $\Delta\phi$ is $0.1°$. With an angular resolution $\Delta\phi$ of $0.5°$ and more it can be detected with difficulty by only four or less measuring points. Thus, an angular resolution of $0.1°$ is pursued for the detection of un-grazed spots on pasture.

Finally, the 2D laser scanner was mounted in the centre and in the front of the vehicle at a fixture consisting of hollow steel profiles. This fixation is rigidly connected with the pivoting front axle. Thus, the laser scanner moves with the axle. Moreover, the scanner is slightly inclined by $2°$ to reduce the distance x_a (see Figure 48) between the contact patches of the front wheels and measurement point on the ground. The reason for this was the use of the connecting line between the two contact patches of the two front wheels as the reference for grass height measurements. Thus, the measurement points have to be as near as possible to this line to reduce measurement errors because of the rough ground surface on pastures.

Δx_a

Figure 48: Position of the laser points in relation to the vehicle

6.5.2 Preliminary studies in regard to soil spot detection

As mentioned above, the output VRE is used for detecting spots on pasture covered with soil. Preliminary studies were performed to analyse the characteristic VRE for grass and soil zones. For this purpose, a grassy area which was covered on one half with soil was scanned to compare the values of the two zones (see Figure 49).

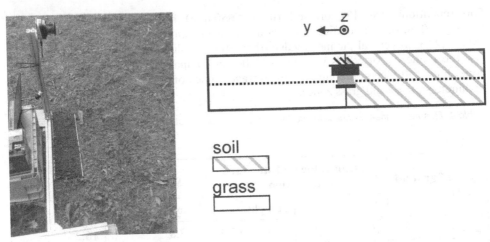

Figure 49: Measurements for analysis of characteristic VRE values of grass and soil zones

The resulting measured values VRE for one laser head rotation and in relation to the y-position, which was calculated by the output distance values d_i of the laser for each measuring point, are shown in the diagram of Figure 50. Table 2 contains the calculated mean values and the coefficient of variation for each zone.

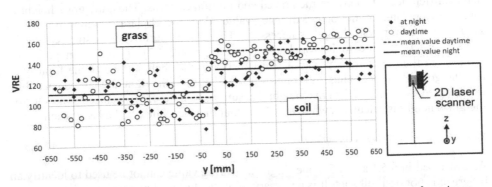

Figure 50: Results of measurements to distinguish soil from grass (measuring values of one laser head rotation)

Without further data analysis the diagram shows higher VREs of the soil zone. Moreover, a slighter scatter of the values within the soil zone can be seen. The detailed values are listed in Table 4. The particular coefficients of variation are 17.8 % (grass) and 7.5 % (soil) according the daytime measurement. At night the difference of the scattering values (soil and grass) are with 5.8 % lower. With 22500 values on each zone (corresponds to 500 laser rotations), on soil (0 < y < 650 mm) the mean value was 148, on grass (-650 < y < 0 mm) it was 106 (daytime measurement). The coefficient of variation for the soil zone was 7.5 % and for grass 17.5 %. At night the mean value of

500 repetitions was 131 on soil $(0 < y < 650$ mm) and on grass it was 111 (-650 $< y < 0$ mm). The coefficient of variation for the soil zone was 11.0 %, for grass it was 15.2 %. In general the mean value of amplitude value of the grass zone was lower and the scattering of VREs was much higher in comparison to the values of the soil zone. Especially the scattering and the average values of VRE values are characteristic parameters of soil and grass zones.

Table 4: Results of analysis according to the detection of soil spots on grass areas

type of ground surface	mean value and standard deviation		coefficient of variation [%]	
	day	at night	day	at night
grass	105.6 ± 18.8	110.7 ± 17.9	17.8	16.2
soil	149.9 ± 11.3	133.3 ± 13.8	7.5	10.4

6.5.3 Algorithms for leftover and soil spot detection

Spots, where mulching is required after grazing on pasture, are detected by the grass height difference Δz_g between the grazed and un-grazed areas. Thus, the grass height z_g must be calculated by the output distance values d_i of the 2D laser scanner. For this, a zero reference for grass height is necessary. The ground surface on pasture cannot be used as reference due to its roughness (see paragraph 5.4.2). Two known points on ground are the two contact points A and B between the shell surface of the front tyres and the ground surface. Thus, the connection line between the right and left contact point can be used as reference for grass height measurement in the first instance. Consequently the grass height is calculated while d is the output distance value of the laser scanner:

$$z_g = z_{laser} - \cos(\beta) \cdot \cos(\varphi) \cdot d$$

(6.5.3.1)

As described in 6.5.1 and 6.5.2 one single measuring value cannot be used to identify an un-grazed spot or a soil spot. It is necessary to use a cluster of measuring points. For this reason the scanning zone is divided into $n = 26$ sections with a width of 50 mm (see Figure 51).

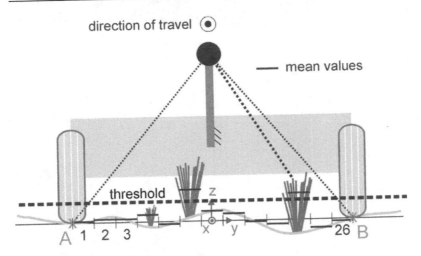

Figure 51: Sketch of spot detection

The algorithm calculates the height z_{gj} of each laser beam and the grass height average $\bar{z}_{gn}$ in each section to analyse the sensor data for one laser head rotation. It checks, if the grass height average $\bar{z}_{gn}$ exceeds a set threshold z_t. Moreover, it counts the number of sections n_e where this is the case. If n_e is higher than a set threshold n_t, the algorithm decides for "un-grazed", else for "grazed". It is expected that both parameters z_t and n_t influence the detection rate because of the rough soil surface on pastures. Both parameters determine the sensitivity of the spot detection or rather they define how easily the mulcher is triggered.

Furthermore the algorithm calculates the average value of VREs in each section and the standard deviation of the measured VREs within a section is calculated. In addition it checks, if the average value is in the range between 120 and 160 as well as, if the standard deviation is in a range of ±14. If so, the algorithm identifies a local damage of the sward in this section. The number of sections with sward damages n_d is counted. If it exceeds a set threshold n_{dt}, a spot where seeding must be applied is identified as such.

Finally, the current status of the required operation (mulching or seeding) is continuously determined.

7 Modeling the machine

7.1 Vehicle model

As defined in chapter 5.4 the machine has to follow predefined paths. Therefore the forward kinematics of the vehicle is of interest in regard to have a calculation model of the trajectory, also known as dead reckoning [58]. The front wheels of the base vehicle (see paragraph 6.1) are steered by a steering linkage which is moved by a double-acting hydraulic cylinder (see Figure 52).

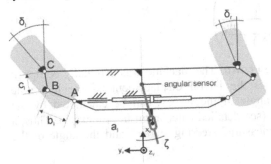

Figure 52: Steering kinematics of the vehicle front wheels

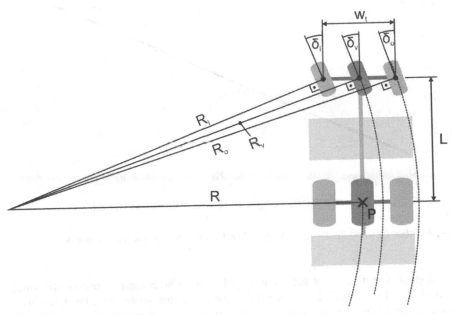

Figure 53: Single track model of the machine

It is assumed that the steering follows the Ackermann rule (Figure 53)[73]. In that case, the vehicle can be reduced to the single-track model. That means that the steering kinematics is reduced to one single virtual central axis [73]. The movement of the point P between the rear wheels is considered for vehicle guidance. In the so-called dead reckoning the side slip angle β plays the leading role. In our case, if only the front wheels are steered, the angle β corresponds to the steering angle of the centre virtual front wheel δ_v [73]. It is the parameter which must be finally controlled. It can be calculated as follows:

$$\beta = \delta_v = arctan\left(\frac{L}{R}\right) = arctan\left(\frac{L}{\frac{L}{tan\,(\delta_i)} + \frac{w_t}{2}}\right) \tag{7.1.1}$$

Thus, the steering angle δ_i or δ_a must be known. The machine was originally equipped with an angular sensor, which measures the lateral displacement of the hydraulic cylinder (see Figure 52). The steering angle of the wheels can be calculated by the known geometry of the steering linkage (see detailed calculation in Annexe 5). Thus, a mathematical relationship between the measured steering angle ζ and the angle of the wheel can be made.

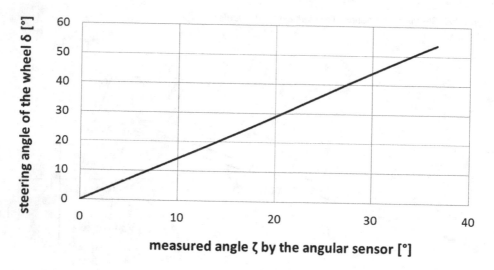

Figure 54: Relation between the measured angle ζ and the real steering angle of the wheel

A high slippage was found for the front wheels during steering because of the small wheels and the low wheel load. For this reason, steady-state circular tests were performed to determine the slippage and get a relationship between the real side slip angle β and the output value of the sensor ζ or rather the voltage U for developing the steering control. The vehicle was equipped with a GPS RTK system. The antenna was attached in the centre of the vehicle at the rollover protection structure with a small displacement of $\Delta x_{GPS} = 110$ mm in front of the rear axle. As mentioned above point P is of interest in regard to vehicle motion. A calculation (see Annexe 6) has shown that this displacement of 110 mm can be neglected.

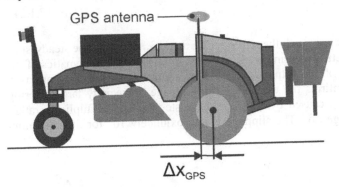

Figure 55: Position of the GPS antenna related to the vehicle

NMEA (National Marine Electronics Association) standard data records are used to detect GPS data. Different circles with constant steering angles were driven with a speed of v = 0.7 m/s (expected operational speed) on a grassland area and the GPS positions as well the angular sensor output voltage U have been recorded. The results are exemplarily shown in Figure 56.

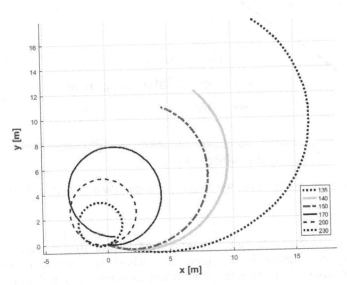

Figuro 56: Resulting circles of steady-state circular tests for different command values

The diameters D of these recorded circles were measured and the resulting real steering angle δ_{vr} has been calculated.

$$\delta_{vr} = arctan\left(\frac{L}{0.5 \cdot D}\right) \tag{7.1.2}$$

The diagram in Figure 57 shows both the real steering angle δ_{vr} based on the steady-state circular tests and the theoretical steering angle δ_v based on the steering kinematics of the steering linkage depending on the output voltage of the angular sensor U. A comparison of both steering angles confirms the expectation of slippage. Especially large steering angles lead to understeering caused by slippage. Low steering angles implicate rather small steering angle slippage $\Delta\delta_v$. The slippage is approximately 10° for the maximum steering angles.

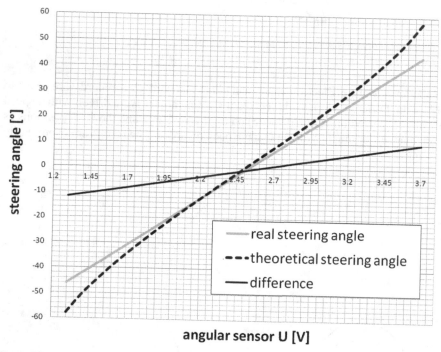

Figure 57: Real steering angle and theoretical steering angle depending on the output voltage of the angular sensor

The driving speed v can be detected directly by recording NMEA data. But additionally the speed and the travelled distance is measured at a wheel for validation. An incremental encoder was installed at the right front wheel to determine the distance travelled and calculate the speed. Wheel slip of the non-driven front wheels was negligible. A circular slab with n_{screw} cylinder head screws around the circumference d_s was fixed at the rotating wheel rim. An inductive sensor at the wheel fork detects the impulses of the passing screws (see Figure 58).

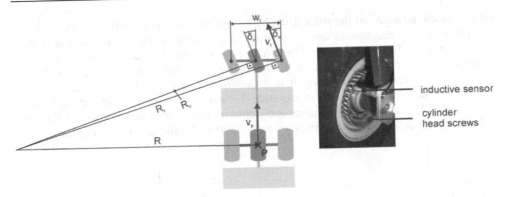

Thus, the counted number of impulses result in the distance travelled d_t and the current speed v_t by taking the wheel diameter and the time between pulses into account:

$$d_t = \frac{\pi \cdot d_w}{n_{screw}} \cdot \sum \Delta n_{si} \qquad (7.1.3)$$

$$v_r = \frac{\pi \cdot d_w}{n_{screw}} \cdot \frac{\Delta n_s}{\Delta t} \qquad (7.1.4)$$

But as mentioned above, the point P in the middle of the vehicle is of interest in terms of vehicle motion. For this reason a conversion of the measured travelled distance of the right front wheel d_t and its speed v_r to the point P is necessary (see detailed calculation in Annexe 7).

$$v_P = \frac{\Delta s_v \cdot v_r}{\Delta s_r} = \frac{R_v}{R_r} \cdot v_r = \frac{sin\delta_r}{tan\delta_v} \cdot v_r \qquad (7.1.5)$$

7.2 Mulching process model

The lowering and raising process of the mulcher and the rotor acceleration are analysed in this section. These functions are of interest in regard to the development of machine control. Especially a mulcher model is necessary for programming the operating mode of selective mulching. If an un-grazed spot on pasture is detected by the 2D laser scanner, the system has to react by starting the mulching process. The same applies to deactivate the mulcher if no left over is detected anymore. First the lowering and lifting process was analysed. One objective was especially to find out the time which is needed to lower and raise the mulcher. Therefore these processes were captured by a high-speed camera. The times when the mulcher had started to move and reached its positions were noted for the lowering and lifting process to get their periods of time Δt_{low} and Δt_{lift}. Measurements of the rotational speed of the rotor drum by an optical measurement device showed that the rotor is fully accelerated during the lowering process if the acceleration is started simultaneously. Thus, the period of time Δt_{low} is considered, because the time needed for acceleration of the rotor drum is shorter. The time for lowering the mulcher is $\Delta t_{low} = 1000\pm30$ ms and for lifting it is $\Delta t_{lift} = 700\pm20$ ms. The following diagrams in Figure 59 show qualitatively the assumed behaviour of the mulcher up and down movement. The curve characteristics are depicted linear. It is not known, if they are linear, but the detailed characteristics are not important for the next considerations.

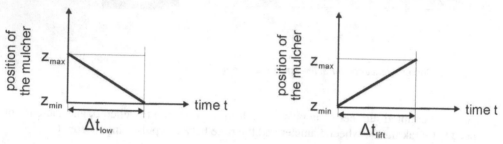

Figure 59: Qualitative time behaviour of lowering and lifting the mulcher

7.3   Seeding process model

First the spreader for applying seeding had to be calibrated for grass seeds to get the relation $\dot{V}(n_{shaft})$ between the seed rate and the seeding shaft speed. In a first step the lateral distribution was analysed (see Figure 60) and, if applicable, was adjusted to guarantee an even distribution.

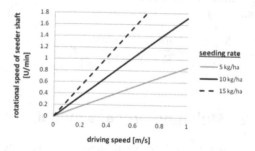

Figure 60: Tests to set an even lateral distribution and to calibrate the seeder

Figure 61: Rotational speed depending on the driving speed v for different seeding rates

The seeding shaft was driven with different speeds in a certain time and the masses of dropped out seeds were weighted afterwards. The seeding shaft speeds is of course dependent on the vehicle speed. Finally, the relation between the seeding shaft and the driving speed for different seed rates can be determined. These linear relations are shown in the diagram of Figure 61.

8   Development of machine control

The machine can execute its missions for automatic pasture maintenance in two modes. One is mulching the whole area, actually the conventional method. The machine follows a GPS trajectory which fully covers a paddock area by parallel tracks with an inter-distance of one working width. The mulcher is switched on and lowered in the beginning of each track and is permanently active in float position. The mulcher is lifted and deactivated only during turning manoeuvres. The new mode in terms of precision farming (see paragraph 3) is selective mulching of un-grazed spots after grazing. Thereby the vehicle also follows a GPS trajectory according to conventional method by parallel tracks. The mulcher is switched on but not lowered until the machine detects an un-grazed spot in the sense of precision farming. The control, especially of the second mode is developed in the following sections.

8.1   Basic system control architecture

In principle, the machine control has three tasks as a result of the requirements described in paragraph 5.4:

- Manual remote control from a supervisor PC
- Automatic vehicle guidance
- Localisation of spots where mulching and seeding operations are required
- Implement control (seeder and mulcher)

The machine is originally radio remote controlled (see layout description in paragraph 6.1). The remote control system, consisting of remote control and receiver unit, is replaced by a new hardware, the so-called "low level controller" (LLC) and the "remote supervisor" (see Figure 62). A myRIO hardware device from National Instruments (NI) [74] combined with a X-CAN Adapter from Stratom [75] is used as LLC for the prototype. This CAN interface was connected to the original machine control (OMC). Thus, it was possible to program the machine control within a LabView environment (NI software) by sending CANopen commands to the OMC in the manner of the original receiver unit. The LLC is connected to remote supervisor, a usual desktop PC over a wireless network. Thus, all original functions of the machine can be controlled manually, remotely e.g. by a commercial gamepad controller, which is connected to this PC. All actuators of the machine including the combustion engine can be controlled based on commands sent from the supervisor PC or directly from the LLC itself. The localisation of spots, where pasture maintenance operations are required, is also implemented into the LLC. For this reason, the 2D laser scanner is connected with this device. Another additional hardware device is the so-called high level controller (HLC). It is used to implement the automatic guidance of the vehicle. In our case, a hardware of the type "Effinav" from the French company Effidence was used. The following Figure 62 shows the architecture of the basic system with the particular devices.

© The Editor(s) (if applicable) and The Author(s), under exclusive license to Springer-Verlag GmbH, DE, part of Springer Nature 2020
B. Seiferth, *Development of a system for selective pasture care by an autonomous mobile machine*, Fortschritte Naturstofftechnik,
https://doi.org/10.1007/978-3-662-61655-0_8

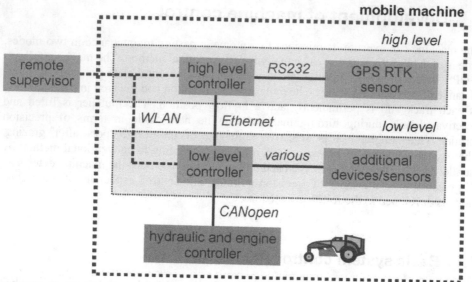

Figure 62: Architecture of the basic system

If the WiFi connection to the Host PC is interrupted, the machine stops automatically. The connection between both the Host PC and the low level controller is continuously monitored by a kind of request-response method. If the Host PC does not send the negation of the value which has been sent by the LLC, the connection is interrupted and the low level controller stops the machine.

The machine is equipped with a 12 V socket which is intended to connect for example a rotating beacon and other additional hardware devices. In order to avoid a voltage drop during the start of the engine an additional battery with a charge controller has to be installed. This battery ensures the maintenance of the voltage level and avoids a power-down of the hardware for this moment.

8.2   Vehicle guidance

After modeling the vehicle (see paragraph 7.1) the control of the automatic vehicle guidance function is developed. First of all, predefined paths, which the machine has to follow (compare 5.4) have to be created. A trajectory planning software for partial and full field coverage has been developed at the French National Research Institute of Science and Technology for Environment and Agriculture (IRSTEA) [76]. The software creates a file with a list consisting of sequent positions (x_{0i}, y_{0i}). The singular positions characterise the reference path, which the machine has to follow. The method of the layout of these trajectory points is described in detail in [76]. Finally, this file is sent to the HLC of the machine at the beginning of its mission for field operations on pasture.

Figure 63 shows important parameters of following the planned trajectory. The reference path is designated with r. It is given by the mentioned file consisting of several points (x_{0i}, y_{0i}). The control variable along the path is s_t. A virtual window with certain dimensions moves with the machine. Points within this window or frame are used to approximate a mathematical function $y_0(x_0)$ of the current part of the trajectory by the least square method of second order (see Figure 63). The curvature $c(s)$ is the curvature of the path at point M, which is the closest one to the path to the point P [77]. Furthermore $\theta_r(s_t)$ is the orientation of the tangent at the point M with respect to the absolute coordinate system.

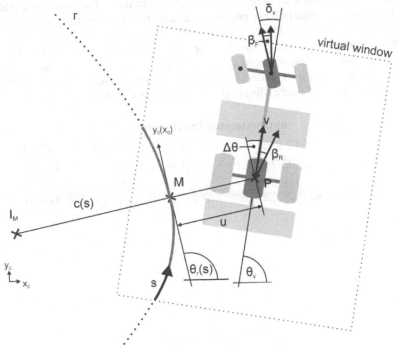

Figure 63: Scheme of automatic guidance of the machine for automated pasture care

In principle, the lateral deviation y and the angular deviation $\tilde{\theta}$ have to be always as low as possible. Figure 64 shows the closed loop system of the vehicle guidance. The high level controller controls the steering angle δ_v so that both parameters are near of zero. The current GPS position (latitude X_{OP}, longitude Y_{OP}) and heading angle Θ_{vm} is detected from the DGPS system. Thus, the lateral deviation y and the crab angle $\tilde{\theta}$ of the vehicle are calculated to be regulated to zero by the high level controller. The resultant output value contents the steering angle δ_v which is sent to the low level controller as command value.

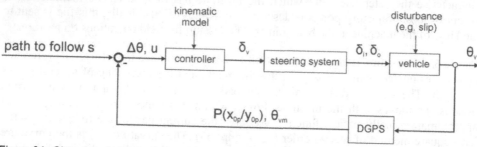

Figure 64: Closed control loop of automatic guidance

A function $\text{cmd}_{\text{steering}}(\delta_v)$ is used to output the steering command. This mathematical relationship is based on the model of the previous paragraph developed by steady-state circular tests (see paragraph 7.1). The algorithms of [77] use the reciprocal of the curvature c, which can be calculated from the steering angle δ_v as based on equation (7.1.1). Finally, the following function $\text{cmd}_{\text{steering}}(1/c)$ was implemented in the low level controller. This function was approximated based on the results of the described steady-state circular tests. It consists of four different polynomial functions f(c):

curvature c [1/m]	approximated function f(c)	
$-0.8 < c < -0.2$	$-164.74\ m^2 \cdot c^2 - 317.14\ m \cdot c + 102.78$	(8.2.1)
$-0.2 < c < 0$	$1095.7\ m^2 \cdot c^2 + 32.91\ m \cdot c + 124$	(8.2.2)
$0 < c < 0.2$	$1046.5\ m^2 \cdot c^2 - 12.44\ m \cdot c + 122$	(8.2.3)
$0.2 < c < 0.8$	$261.91\ m^2 \cdot c^2 - 398.26\ m \cdot c + 151.26$	(8.2.4)

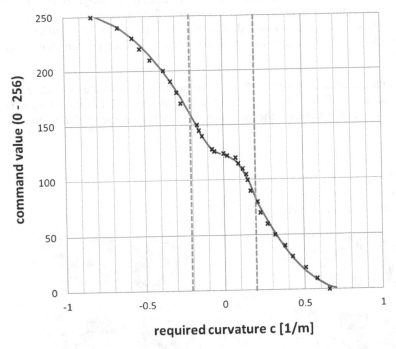

Figure 65: Mathematical relationship between command value and required curvature

8.3   Mulcher control

The development of the mulcher control is described in this section, especially the interaction with the spot sensing system (see paragraph 6.5) in regard to selective mulching is explained. As mentioned in chapter 6.5.3, the localisation system including the 2D laser scanner outputs continuously the actual required status (mulch or not mulch) during the mission. The mulcher control has to react accordingly. Therefore the timing must be considered with regard to the kinematics of the mulcher. Then (in chapter 8.3.2) the kinematics of selective mulching is energetically analysed in theory to develop an energy-efficient mulcher control.

8.3.1 Consideration of timing based on mulcher kinematics

Selective mulching mode means the mulcher is activated and lowered on un-grazed spots. If the machine passes over such a detected spot, the mulcher rotor is engaged and lowered. The mulcher rotor is deactivated and the mulcher is lifted when the spot is mulched. It would be also possible to waive lifting the mulcher and only control the mulcher rotor, but raising the mulcher, if it is not active, has more benefits in regard to sward protection. Thus, the next considerations focus on selective mulching with raising the mulcher. In this context, the period of time between detecting an un-grazed spot and the moment, when the mulcher is lowered and the rotor is fully accelerated, is decisive. This period mainly consists of a measured latency time of control $\Delta t_L = 20$ ms and finally the time Δt_{low} for lowering the mulcher including accelerating the rotor (see paragraph 7.2).

Different cases with different distances between two un-grazed spots are considered in the following. The cases differ in the moment when the second un-grazed spot is detected respectively to the state of the mulcher. The states of the mulcher are the following:

- lowered and active (1)
- in elevation, reducing rotor speed (2)
- lifted and not active (3)
- rotor acceleration and lowering. (4)

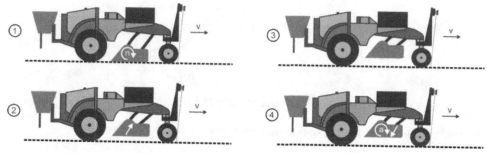

Figure 66: Different states of the mulcher depending on its position and operation

The first considered case is shown in Figure 67. The diagram contains the signal of the spot localisation and the position of the mulcher depending on the time during this scenario.

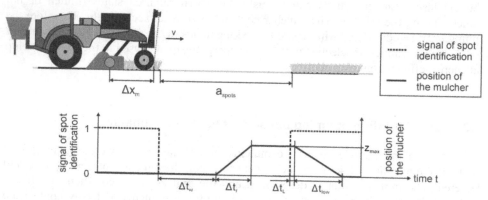

Figure 67: Case 1a - a next un-grazed is detected during the mulcher is in its maximal position z_{max}

The vehicle passes over an edge of an un-grazed spot to a grazed area. The mulcher is first lowered and active. When the signal changes from 1 to 0, there is not an un-grazed spot anymore and the mulcher has to remain switched on and lowered for the time Δt_w to guarantee mulching spots completely. It depends on the distance $\Delta x_m = 0.8$ m between the measurement point of the laser scanner and the centre of the mulcher rotor (see Figure 67 above) and the machine speed of $v = 0.58$ m/s (see paragraph 6.1 and equation (8.3.1.1)).

$$\Delta t_w = \frac{\Delta x_m}{v} \qquad\qquad (8.3.1.1)$$

After this time Δt_w the mulcher is stopped and lifted to its maximum position z_{max}. Then after a while the vehicle passes over an edge of a following un-grazed spot and the signal changes from 0 to 1. After the time of latency Δt_L, the activation procedure of the mulcher starts and the mulcher is lowered. It is fully lowered after the time Δt_{low}. Figure 68 shows the situation when the second un-grazed spot is detected during the mulcher is being lifted. The signal changes from 0 to 1 and after the latency time Δt_L the lifting process is being interrupted and the mulcher is being lowered and accelerated again.

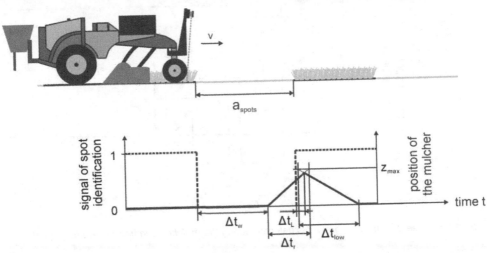

Figure 68: Case 1b - a next un-grazed is detected during the mulcher is being lifted

Both described cases result in a maximum operational speed according the following condition:

$$v_{max} < \frac{x_m}{\Delta t_L + \Delta t_{low}} \qquad\qquad (8.3.1.2)$$

If this condition is fulfilled, a complete mulching of consecutive spots in a distance of

$$a_{spots} > v \cdot \Delta t_w \qquad\qquad (8.3.1.3)$$

is guaranteed. In our case the maximal speed is $v_{max} = 0.66$ m/s. The required operational speed is with 0.5 m/s (see paragraph 6.1) below this value. Consequently selective mulching by the described procedure is possible for the described case. The speed can be higher, if the maximum lifting height z_{max} or the required time for lifting Δt_{lift} is reduced.

The next case shows the situation when the distance a_{spots} between consecutive un-grazed spots is shorter than in the previous case:

$$0 < a_{spots} < v \cdot \Delta t_w$$

$$(8.3.1.4)$$

This time, the moment the second un-grazed spot is being detected during the mulcher is still fully active to complete mulching the first spot (see diagram in Figure 69).

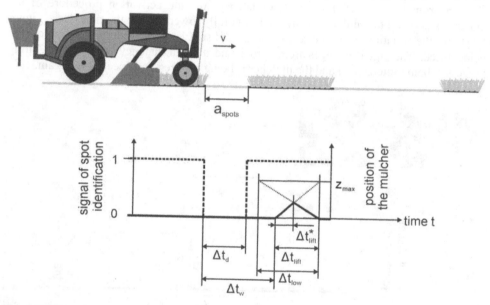

Figure 69: Case 2 - a next un-grazed is detected during the mulcher is in floating position ($z \approx 0$)

The distance a_{spots} is known before the mulcher has finished machining the first spot. Thus, the control can disengage the mulcher so that the second spot can be mulched timely. The time for lifting the mulcher after having finished mulching the first spot and until the mulcher must latest be lowered is defined as follows neglecting the time of latency (see Annexe 8):

$$\Delta t^*_{lift} = \frac{\Delta t_d}{\left(1 + \frac{\Delta t_{low}}{\Delta t_{lift}}\right)}$$

$$(8.3.1.5)$$

It is expected, that the signal from the spot localisation changes quite frequently because of the conditions on pastures. Un-grazed spots consist of blades of grass so that the grass height is not constant within one spot. This circumstance can lead to a frequent deactivation and activation of the mulcher, which must be avoided. Consequently certain filter delay is integrated in the mulcher control software so that the mulcher is at least active for defined 250 mm travelled distance after releasing the signal for mulching.

8.3.2 Energy-optimised mulcher control

In this chapter, the energy requirement of the mulching operation is theoretically considered. Selective mulching is compared with mulching the whole area. It was expected, that the mode of selective mulching has a potential to fuel savings. The conditions for fuel savings are analysed. For this, the power requirement must be examined in detail for each mode. First the power requirements of both modes are calculated separately and then compared.

The Figure 70 shows scenarios of mulching, which are the results of consideration in the preceding paragraph 8.3.1 and which are considered to analyse the energy requirement. The mulcher position as well the grass condition are depicted and put into relation with a diagram showing the qualitative sequence of power requirement of the machine depending on the time t.

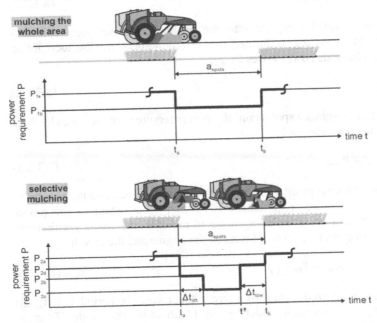

Figure 70: Scenarios of mulching

The first one illustrates mulching the whole area. There are two zones of power requirements P_{1a} and P_{1b}. If there is an un-grazed spot, the power requirement result from the vehicle for locomotion P_{loc_ml} including lowered mulcher, the active mulcher under load P_{m_fl} and power losses P_{loss_1a} of the engine and the drivelines.

$$P_{1a} = P_{m\_fl} + P_{loc\_mlow} + P_{loss\_1a} \qquad\qquad (8.3.2.1)$$

If the machine with active mulcher passes a grazed zone, the power requirement of the mulcher decreases to P_{m_pl} because grass material is not processed. Thus, the total power requirement is P_{1b}.

$$P_{1b} = P_{m\_pl} + P_{loc\_mlow} + P_{loss\_1b} \qquad\qquad (8.3.2.2)$$

If the machine passes over a next un-grazed spot, the power requirement decreases again to P_{1a}.

The next draft of Figure 70 illustrates the mode of selective mulching. At the beginning the power requirement is as high as in the first scenario P_{2a} while mulching an un-grazed spot. If the machine passes over an edge of this spot, the mulcher rotor slows down and the mulcher is raised leading to a power requirement P_{lift_m}. Now the implement is lifted and not anymore on the ground. Thus, the power requirement for locomotion changes to P_{loc_m+v}. The power requirement P_{2b} is as follows:

$$P_{2b} = P_{loc\_m+v} + P_{lift\_m} + P_{loss\_2b} \qquad\qquad (8.3.2.3)$$

If the mulcher has been lifted completely to the position z_{max} (see paragraph 8.3.1 case 1), the power requirement P_{2c} is limited to the requirement for the locomotion P_{loc_m+v} with lifted mulcher plus power losses P_{loss_2c}:

$$P_{2c} = P_{loc\_m+v} + P_{loss} \qquad\qquad (8.3.2.4)$$

If the machine detects an un-grazed spot again, the power requirement increases by P_{a_m} because the rotor has to be accelerated.

$$P_{2d} = P_{loc\_m+v} + P_{a\_m} + P_{loss\_2d} \qquad\qquad (8.3.2.5)$$

If the mulcher reaches its lowest position, the rotor is fully accelerated and the un-grazed spot is mulched. Now the power requirement is again P_{1a} consisting of the power requirement for vehicle locomotion including lowered mulcher, the power requirement for machining the grass material and power losses of the engine and the drivelines.

$$P_{2a} = P_{1a} = P_{m\_fl} + P_{loc\_mlow} + P_{loss\_1a} \qquad\qquad (8.3.2.6)$$

Finally, the gap between un-grazed spots is of interest (that means the period $\Delta t = [t_a, t_b]$) because the power requirements while mulching are identical in both modes. Energy-savings of one mode depend on the amount of energy requirement of each mode in this interval. Thus, the condition for a higher effectiveness of selective mulching according energy requirement compared to mulching the whole area can be expressed as follows:

$$\int_{t_a}^{t_b} \sum P_{1i}\, dt > \int_{t_a}^{t_b} \sum P_{2i}\, dt \qquad\qquad (8.3.2.7)$$

It is not possible to get all machine-specific parameters without great effort. For this reason, stationary measurements of fuel consumption have been carried out to get directly the particular fuel consumption $\dot{V}_i$ which are related to the necessary and relevant power consumptions P_{1a}, P_{1b}, P_{2a}, P_{2b}, P_{2c} and P_{2d} excluding the power requirements for movement P_{loc_mlow} and P_{loc_m+v}. It is assumed, that both requirements for movement of the vehicle with lifted and mulcher in float position are identical. The machine was stationary and performed different actions for 15 min while the engine speed was maximal. The extra tank (see paragraph 6.1) was weighed before and after. The consumption was measured for the engine with no load and full throttle position to get the fuel consumptions $\dot{V}_{1b}$ (running mulcher without load) and $\dot{V}_{2c}$ (engine at idle). Moreover, a control cycle with a repeating lifting and lowering of the mulcher was programmed to measure and calculate the fuel consumption $\dot{V}_{2b}$. Another test cycle contained a repeating accelerating of the mulcher rotor. Thus, the fuel consumption $\dot{V}_{2b}$ and $\dot{V}_{2d}$ were determined pro rata temporis. The following figure shows the results of these stationary measurements.

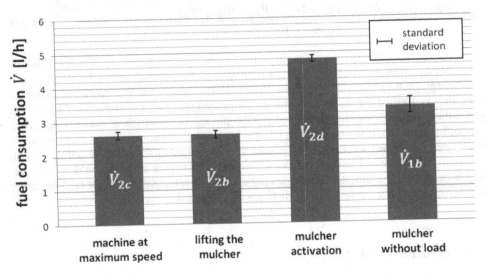

Figure 71: Results of stationary measurements of fuel consumption

These results show that there is hardly any difference between the fuel consumption with engine at maximum speed and lifting additionally the mulcher. During mulcher activation (rotor acceleration and lowering of the mulcher) the fuel consumption increases to (4.8 ± 0.10) l/h. If the mulcher rotates without load, the fuel consumption is (3.4 ± 0.23) l/h.

Now the fuel consumptions of the described scenarios (see Figure 70 on page 67) can be calculated and compared. As mentioned, the part which is of interest is the part where the grass is grazed $\Delta t = [t_a, t_b]$, because the power requirements or rather fuel consumption while mulching are identical in both modes. Fuel-savings of one mode depend on the amount of fuel consumption of each mode in this interval. Thus, the effectiveness of selective mulching according energy requirement compared to mulching the whole area is higher, if the amount of the integral over this period of the sum of fuel consumptions $\dot{V}_{1i}$ is larger than the amount of the integral of the sum of fuel consumptions $\dot{V}_{2i}$ analogous to equation (8.3.2.7):

$$\int_{t_a}^{t_b} \sum \dot{V}_{1i}\, dt > \int_{t_a}^{t_b} \sum \dot{V}_{2i}\, dt \qquad\qquad (8.3.2.8)$$

The first and the second scenario of Figure 70 are compared. The time interval $(t_b - t_a)$ and therefore, the distance between un-grazed spots a_{spots} has certainly influence on the energy requirement of the machine operation mode, besides the machine-dependent parameters. The time t_b depending on the measured fuel consumptions $\dot{V}_{1i}$ and $\dot{V}_{2i}$ of the machine can be calculated as follows (detailed calculation see Annexe 9):

$$t_b > \frac{\Delta t_{lift} \cdot \dot{V}_{2b} - \Delta t_{low} \cdot \dot{V}_{2c} - \Delta t_{lift} \cdot \dot{V}_{2c} + \Delta t_{low} \cdot \dot{V}_{2d}}{(\dot{V}_{1b} - \dot{V}_{2c})} \qquad\qquad (8.3.2.9)$$

Substituting the measured values into this equation the result is $t_b = 2.8$ s which corresponds to a distance of $a_{spots} = 1.6$ m if the operational speed is $v = 0.58$ m/s (see paragraph 6.1). Finally, a fuel-saving effect cannot be expected, if selective mulching is applied for distances between un-grazed spots a_{spots} lower than 1.6 m. That particularly applies for the case 2 of paragraph 8.3.1, in which a fuel-saving effect cannot be expected. It is important to point out, that the fuel consumption measurements have been performed with maximal engine speed. If the engine speed is reduced to an optimum, the distance between the conditions for fuel-saving effects would be smaller, which is better. The time t_b for fuel-saving effects by selective mulching will be lower and consequently the corresponding distance of un-grazed spots a_{spots} as well.

8.4   Seeder control

The control of the seeder is more trivial than the control of the mulcher because there is not an up and downward movement and the acceleration time Δt_a of the seeding shaft is much lower than 0.1 s. The distance between the measurement point of the laser scanner and the downpipes of the seeder is called x_s. Due to Δx_s is 3300 mm, there will not be problems with timing in regard to the short acceleration time of the seeder shaft and the operational driving speed of 0.58 m/s (see paragraph 6.1). Thus, when the signal of spot detection changes from 0 to 1, the control has to activate the seeder after the time Δt_{s_a}

considering the time delay because of the distance x_s minus the time needed for the seeding shaft start-up Δt_{sh_a} and the fall time Δt_f of the seeds depending on the drop height h_f.

$$\Delta t_{s\_a} = \frac{x_s}{v} - \Delta t_{sh\_a} - \Delta t_f \qquad (8.4.1)$$

When the signal of spot detection changes from 1 to 0, the seeder has to remain analogously active for a time Δt_{s_da}:

$$\Delta t_{s\_da} = \frac{x_s}{v} - \Delta t_{sh\_da} - \Delta t_f \qquad (8.4.2)$$

The time delay because of the distance x_s, the time needed to stop the seeding shaft Δt_{sh_da} and the fall time Δt_f of the seeds depending on the drop height h_f have to be considered. The diagram in Figure 72 shows the detection and the control signals depending on the time if the machine passes over spots with an length of l_s where seeding is required.

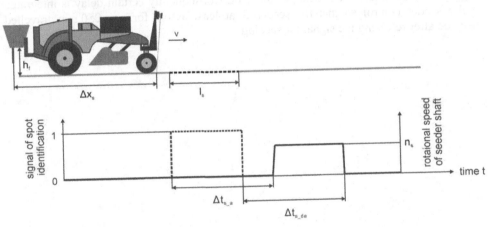

Figure 72: Signal paths during seeding of one spot with a length of l_s

Finally, the seeder control has to trigger the stepper motor of the seeding shaft by a clock signal (see paragraph 6.4) which results in the desired seed flow Q_s depending on the set seeding rate, the driving speed v and the timing of the detection signal. Figure 73 shows the scheme of the seeder control.

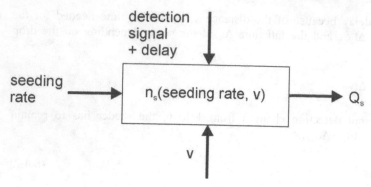

Figure 73: Scheme of seeding control (open-loop control)

Spots where seeding is required are often pervaded with grass. This circumstance can lead to a high frequent deactivation and activation of the seeder, which must be avoided to guarantee definitely the seeding of the spot. Consequently certain delay is integrated in the seeder control so that the seeder is at least active for e.g. 250 mm travelled distance after releasing the signal for seeding.

9 Evaluation of the developed machine under real conditions

The developed machine has been tested under real conditions on flat pasture areas of the Weihenstephan-Triesdorf University of Applied Sciences near of Freising. The following paragraphs describe the final tests and their results. In general, the functionality of the machine has been tested. The focus is on the localisation of spots where maintenance operation is necessary. Moreover, the expected economic benefit in terms of fuel savings by selective mulching is evaluated. Both issues are analysed with the results of measurement during these field tests.

9.1 Vehicle guidance

First a trajectory is generated by the above-mentioned software (see paragraph 8.2). The generated trajectory covers an area of 40 by 10 m separated into five parallel tracks (see Figure 74) and is loaded into the controller. The lateral deviation at the start point, where the vehicle has been manually manoeuvred to, was 1.5 m. Then the automatic guidance was started. Thus, the guidance with a deviation at the beginning was tested. During the automatic drive the positions x and y, the lateral deviation u of the point P (compare Figure 63) as well as the angular deviation $\tilde{\theta}$ have been recorded.

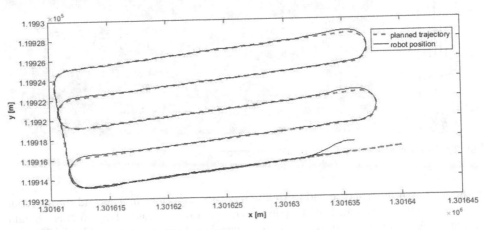

Figure 74: Test of vehicle navigation and guidance

The Figure 74 shows that the vehicle correctly follows the whole trajectory within a deviation of ±20 cm at a velocity of 0.75 m/s. The lateral deviation on straight lines is less than 10 cm. In curves with large steering angles (>30°) it is maximal 30 cm. The half-turn maneuvers are performed with steering angles of maximum 50°.

© The Editor(s) (if applicable) and The Author(s), under exclusive license to Springer-Verlag GmbH, DE, part of Springer Nature 2020
B. Seiferth, *Development of a system for selective pasture care by an autonomous mobile machine*, Fortschritte Naturstofftechnik,
https://doi.org/10.1007/978-3-662-61655-0_9

9.2 Mulching and seeding process

The localisation of un-grazed spots where mulching is required, has been tested. As described in paragraph 6.5.3, the two parameters z_t (threshold for average grass height) and n_t (threshold for number of sections where the grass height average $\bar{h}_n$ exceeds a set threshold z_t) define the detection rate of un-grazed spots on pastures. It was expected (see paragraph 6.5.3) that both values determine the sensitivity of the detection because of the rough ground surface. The smaller both values, the higher is the sensitivity of detecting un-grazed spots. In order to get a reference with known positions of potential spots with high grass, an area has been prepared for test. The machine had to follow a predefined trajectory which fully covered an un-grazed area of a real pasture of nearly 40 m by 8 m separated into eight parallel tracks. The active mulcher was randomly deactivated for a few centimeters by remote control while the positions x_{i_1}, y_{j_1} and the mulcher status S_{m_1} (on = 1/off = 0) have been recorded. Consequently a simulated grazed paddock area with "un-grazed spots" has been obtained (see Figure 75).

Figure 75: Prepared test area with simulated un-grazed spots

In the next step the machine had to follow the same trajectory in a second run to scout the area with not mulched spots (simulated un-grazed areas). The 26 grass height average values of the 50 mm wide sections over the whole scanning width of 1300 mm for all laser head rotations (see paragraph 6.5.3) as well the corresponding vehicle positions x_{i_2} and y_{i_2} have been continuously recorded. This raw data has been used to examine the influence of the two above-mentioned parameters on the sensitivity by post-processing. The evaluation program uses the in 6.5.3 described algorithm and identified un-grazed spots of the second run for different value pairs of the two parameters n_{t_i} and z_{t_j} ($z_{t\_j} = [0; 200]$; $n_{t\_i} = [0; 26]$). The particular average grass height $\bar{z}_{gn}$ of each section and laser head rotation is calculated and compared with the respective set grass height threshold z_{t_i}. The number of sections n_e, where $\bar{z}_{gn}$ exceeds z_{t_i}, is determined and compared with the respective set threshold n_{t_i} (compare 6.5.3). Thus, the result was a virtual mulcher status S_{m_2} (on = 1/off = 0) for each vehicle position x_{i_2} and y_{i_2}. Both the recorded mulcher status S_{m_1} during the run for preparing the field conditions and the second run for scouting S_{m_2} were compared for each position x_i and y_j and different value pairs of n_{t_i} and z_{t_i}. Finally, the number of wrong decisions or rather the percentages of not detected un-grazed spots related to all positions with un-grazed spots n_{ns} and misinterpreted un-grazed spots referred to all positions where the mulcher was active n_{ms} for the whole run and each value pair have been determined. Figure 76 shows points p_j [n_{ms_i}; n_{ns_i}] which represent the number of wrong decisions or rather error rate respectively for one setting according the two threshold parameters z_t and n_t.

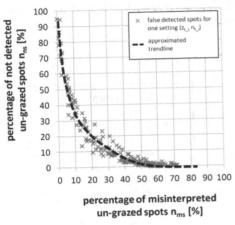 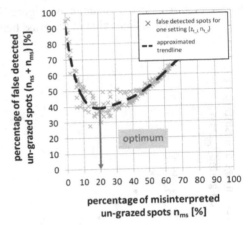

Figure 76: Conflict of spot detection

Figure 77: Diagram to determine the optimal setting according spot localisation

This result confirms the assumption of an influence of the two parameters n_{t_i} and z_{t_j} on the detection rate or rather sensitivity of the localisation system under real conditions. In principle, the percentage of misinterpreted un-grazed spots n_{ms} characterises the sensitivity of the mulcher activation. A higher percentage n_{ms} means that the machine triggers the mulcher more sensitively. The more sensible the detection the more misinterpretations is caused by occurrence of bumps in the ground. The diagram in Figure 76 highlights this conflict according to an optimisation of the spot detection. Either all un-grazed spots are localised, but also grazed zones are misinterpreted or some areas with leftovers are not identified at all.

As a result the sensitivity can be set in both directions. Either the parameter z_t and n_t are set low and all un-grazed spots are localized and mulched, but also grazed zones are misinterpreted. Or both parameters are set high and some areas with un-grazed spots are not identified at all and thus, not mulched. Figure 77 shows that the total optimum results in an error detection rate of about 20 %. As mentioned in section 1.3, the area of un-grazed spots can reach a percentage of up to 10 to 20 % of total pasture area. Thus, 20 % of not detected un-grazed spots (error rate) means an area of 2 to 4 % of the total paddock area. This is contrasted to 80 % of detected un-grazed spots, which corresponds to an area of 8 to 16 % of the total paddock area. If the whole grazing area per season is 240 ha (100 LU, 6 rotations, 0.02 ha/LU, total paddock area of 40 ha), the detected and thus the avoided feed loss area of un-grazed spots is between 19.2 to 38.4 ha, which corresponds to a grazing area for eight to 16 LU per season. Finally, automated pasture care is even profitable with this error rate after few seasons depending on the price situation. If the sensitivity is increased, the error of not detected un-grazed spot can be further reduced.

A closer look at the influence of the particular parameter z_t and n_t shows that the error rate according misinterpreted un-grazed spots n_{ms} declines with an increasing threshold for the number of sections where the set grass height threshold z_{t_j} is exceeded. This trend is shown in the diagram of Figure 78. Moreover, the number of n_{ms} declines generally with a higher value of the grass height threshold z_{t_i}.

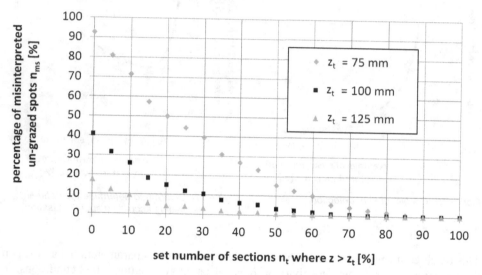

Figure 78: Error rate depending on n_t

Similarly the number of not detected un-grazed spots increases if the thresholds z_{t_j} and n_{t_i} increase (see Figure 79).

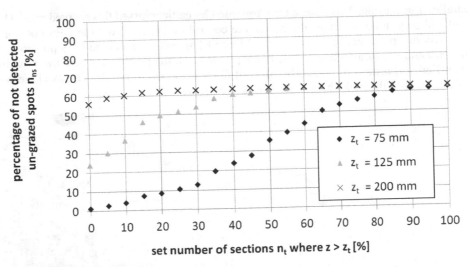

Figure 79: Error rate depending on n_t

The diagram in Figure 80 shows the result of an exemplary insensitive detection of un-grazed spots. The black points mark the positions where the mulcher was inactive during the first run, the white ones mark the positions where not mulched spots have been detected during the second passage. The not mulched zones are detected but not completely. A percentage of 33 % of the not mulched area on the test field has not been detected. The zones, where leftovers are detected, although the mulcher has been active before, are single points.

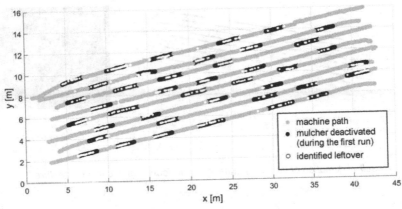

Figure 80: Representative result of a field test for analysis of leftover detection

Next the seeding spot localisation of the machine is analysed. Therefore the machine followed a trajectory which fully covered an area of nearly 40 m by 8 m with eight parallel tracks again. This time, a trail, trampled by cattle, crossed this scouted area. This trail had been followed manually by remote control before to record the GPS positions. The driving speed was (0.6 ± 0.05) m/s. Figure 81 shows the recorded GPS positions of the trail and the parallel tracks. The black points mark the positions, where the machine identified soil instead of grass. The route of the trail is clearly visible. The machine was able to localise the trail where seeding was necessary.

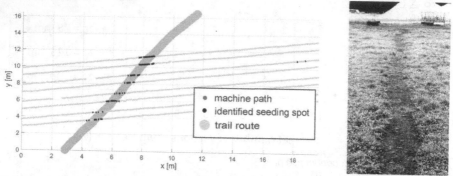

Figure 81: Exemplary results of a field test in order for analysing seeding spot detection by the developed machine (left); cattle trail, which crossed the test area (right)

Moreover, another area has been scouted by the machine. This time it had to scout an area where the sward was damaged around the water trough of one paddock. Before the GPS positions of the trails to the water trough and the boundary of the area with soil have been recorded. These positions comply with the detected areas. Figure 82 shows the result.

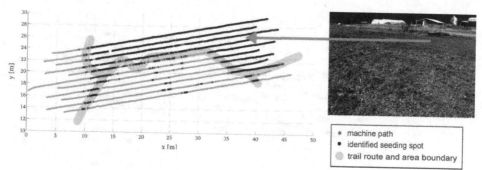

Figure 82: Result of detection of damages caused by the footsteps of cattle on pasture

These test results confirm that the characteristic criteria of soil spots (see paragraph 6.5.3) are sufficient to identify them for seeding actions on pasture.

9.3   Analysis of economy and productivity

The fuel consumption has been measured to compare selective mulching with mulching the whole area under real conditions. For the measurements the machine had to mulch areas of a size of 320 m² with different number of not mulched areas per track n_{spots}. Thus, it followed a predefined trajectory which fully covered these areas of nearly 40 m by 8 m separated into eight parallel tracks (similar to Figure 75). In order to get simulated un-grazed spots with different distances a_{spots} per area, different test areas have been mulched in exception of spots. Thus, areas with different distances a_{spots} of (5 ± 0.05) m, (4 ± 0.05) m, (2 ± 0.05) m and (1 ± 0.05) m have been prepared. The installed extra Diesel tank (see paragraph 6.1) was weighed before and after each measurement run to determine the consumed fuel. The average operational speed v during these measurements was 0.7 m/s resulting in an area performance of 0.91 m²/s. The engine speed was set at maximum. The diagram in Figure 83 shows the measured fuel consumptions of selective mulching and mulching the whole area with different distances of spots a_{spots}.

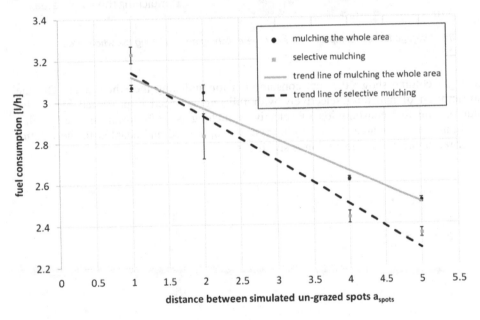

Figure 83: Result of fuel consumption measurements

As expected, the fuel consumption decreases with longer distances a_{spots}. The results show, that selective mulching is more fuel-efficient, if the distance a_{spots} between the un-grazed spots is large $(> 1.5$ m). But for shorter distances a_{spots} $(< 1.5$ m) the fuel consumption of selective mulching is higher compared to mulching the whole area. This is because of the high power requirement for accelerating the mulcher rotor (see Figure 71). The result of this test is similar to the in section 8.3.2 calculated minimal distance a_{spots} of more than 1.6 m which results in a fuel-saving effect by selective mulching.

Further measurements have been performed on 12 pasture areas in each mode with full enigne speed. The averaged percentage of mulching spot area (particular spot area of 1 m²) was (5.4 ± 0.9) %. The average fuel consumptions of these measurements are shown in the diagram of Figure 84.

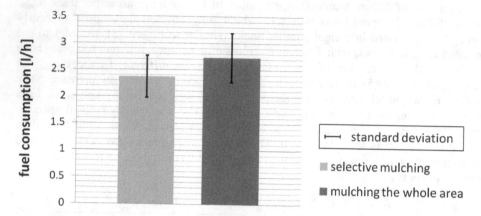

Figure 84: Average fuel consumptions for selective mulching and mulching the whole area

The right column shows the fuel consumption for mulching the whole area. The fuel consumption of mulching selectively with full engine speed is presented by the left column. The fuel consumption of selective mulching is 13 % lower in average than mulching the whole area. If the engine speed is optimised and adapted to the existing power requirement this fuel saving is still higher.

10 Summary and outlook

The degree of automation is still quite low in the outdoor sector such as pasture grazing. There are a few approaches, but the breakthrough has not been occurred yet, as it has happened in animal housing on modern dairy farms. Pasture maintenance is until today much manual and hard work so that it is often waived. Thus, an approach for automation is evident and regular pasture maintenance can even have cost-benefit effects, if the associated additional costs are lower compared with conventional pasture care and lower than the related added yield.

The presented prototype development shows the potential to introduce robotics in pasture maintenance based on the fusion of already existing technologies. A commercially available remotely controlled mulcher has been upgraded, whereby the software of the original machine has not been changed. The prototype has been tested under real pasture conditions. It was able to follow accurately previously planned trajectories based on GPS navigation. An interruption of the GPS signal under trees has been observed during the tests on pasture areas. This limitation can be removed by the use of additional sensors for navigation (see paragraph 2.4).

The machine was able to localise pasture maintenance spots in real-time and perform the appropriate operation directly. Thereby the detection rate error in regard to the localisation of un-grazed spots is about 20 %. The main reason for this is the rough surface on pasture with bumps on the ground. The localisation by the 2D laser scanner is not able to differ between high grass spots and bumps because the reference is the connecting line between the two contact patches of the two front wheels. The tests have also shown that the localisation of sward damages caused by grazing livestock works very well using the 2D laser scanner and its output VRE values both at night and by day.

A fuel-saving effect by selective mulching effect was expected. It can be achieved up to a certain distance between spots a_{spots}. If the distance between spots a_{spots} is very short, selective mulching is more fuel-intensive. But regular pasture care leads to less un-grazed spots, so that the distance between serial spots a_{spots} will be reduced over time and therefore the fuel-saving potential of selective mulching will increase, too. Automated selective mulching has a gentle effect on the sward compared to mulching the whole area with standard machinery. It enables selective pasture maintenance operations like mulching and seeding in terms of precision farming. The considerate handling of the resource pasture and the potential energy saving are advantages of selective pasture maintenance.

Figure 85: The developed prototype for automated pasture maintenance operations

This thesis shows future potentials for automated intelligent pasture maintenance and management. Further potential concepts for a higher efficiency can be deduced from this thesis (see paragraph 5.2). The operation of the mobile machine for automatic pasture care in combination with the use of an UAV can increase the performance of such a system.

Additionally, information about the state of grazing areas is received and the farmer can be assisted considering management strategies. Maintenance spots can be mapped and memorised to optimise pasture management. Thus, the application not only reduces workload, but it also provides useful information about the state of the paddock in terms of "Big Data". The farmer is supported in taking decisions on cattle rotations, stocking rate or renewing or fertilising an area based on this information. Moreover, the grass height principally is measured by the 2D laser scanner. Thus, it is also possible to determine the biomass on pastures because the grass height correlates with the biomass [78]. In this context, an adjusting of the driving speed depending on the biomass, which has to be mulched, as a supplement or alternative to the feedback-control using the load at the rotor is also conceivable for future work. Knowing the biomass on the paddock is helpful in regard to supplement feeding or also to the decision according stocking rate and time of cattle rotation.

The safety aspect has not been considered in this thesis. But it is an important point. In the first instance the automatic pasture maintenance operations are limited to rotational systems. Thus, risks of conflicts between cattle and the machine are avoided. But nevertheless a safety system to detect living being and obstacles to stop the machine if required is absolutely necessary. Scientists have already focused on this topic (e.g. [46]).

Appendix

Annexe 1

Calculation of necessary ground clearance based on detected soil profile (chapter 5.4.2)

All possible pairs of points for different track gauges w_{ti} and wheelbases w_i are analysed and each necessary ground clearance b_i is calculated on the following principle:

Figure A1.1: Sketch of soil profile

First all possible pairs of point for different track gauges w_{ti} and wheelbases L_i are determined. Then each necessary ground clearance b_i is calculated by determining the two linear equations $a(x)$ and $c(x)$ as follows.

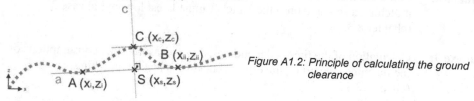

Figure A1.2: Principle of calculating the ground clearance

Linear equations:

$$a(x) = m_a \cdot x + t_a \tag{A1.1}$$

$$\text{with } m_a = \frac{(z_{ii}-z_i)}{(x_{ii}-x_i)} \text{ and } t_a = z_i - m_a \cdot x_i \tag{A1.2}$$

$$c(x) = m_c \cdot x + t_c \tag{A1.3}$$

$$\text{with } m_c = -\frac{1}{m_a} \text{ and } t_c = z_c - m_c \cdot x_c \tag{A1.4}$$

After the point of intersection S is calculated:

$$x_S = \frac{(t_c - t_a)}{(m_a - m_c)} \tag{A1.5}$$

$$z_S = m_a \cdot x_S + t_a \tag{A1.6}$$

Finally, the particular necessary ground clearance corresponds to the distance between the point C and S. The necessary ground clearance results in the following equation:

$$\overrightarrow{CS} = b_i = \sqrt{(z_C - z_S)^2 + (x_C - x_S)^2} \tag{A1.7}$$

Annexe 2

Calculation of the operational time Δt₀ (chapter 6.1)

specific fuel consumption

$$b_e = 240\ g/kWh$$

power

$$P = 18\ kW$$

tank capacity [29]

$$V_f = 18\ l$$

densitiy of Diesel

$$\rho_{Diesel} = 0.84\ kg/l$$

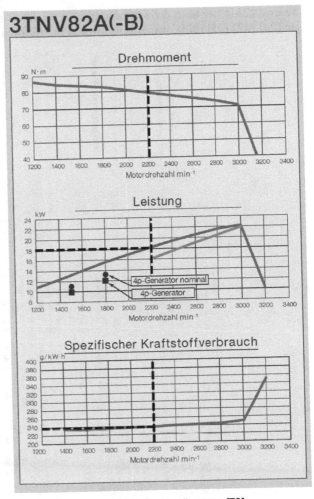

Figure A2.1: Engine diagrams [79]

→ operational time of the machine:

$$\Delta t_o = \frac{V_f}{\frac{b_e}{\rho_{Diesel}} \cdot P} = \frac{18\ l}{\frac{240\ g/kWh}{0.84\ \frac{kg}{l}} \cdot 18\ kW} = 3.5\ h \tag{A2.1}$$

Annexe 3

Evaluation of mulcher types (chapter 6.3)

The evaluation has been done in tabular form based on statements of technical literature (see [24] [22] [25] [80]), which are shown in the following tables:

criteria	weighting factor	scoring points				
		0	1	2	3	4
power requirement	0.05	>20 kW/m$_{AB}$	15-25 kW/m$_{AB}$	10-15 kW/m$_{AB}$	5-10 kW/m$_{AB}$	<5 kW/m$_{AB}$
operational reliability	0.15	Foreign matters (e.g. rocks) cause high damage. The functionality is not guaranteed in case of collision. The machine tends to be blocked by grass material.	The cutting tools are highly sensitive to foreign matters. The functionality is highly restricted in case of contact with foreign matters.	A damage of the cutting tools affects the functionality.	The susceptibility to foreign matters is low. A damage of the cutting tools by foreign matters will hardly affect the functionality.	A safe operation is always guaranteed. The cutting tools do not block even if the conditions are difficult. Foreign matters do not damage the advice.
maintenance effort/wear	0.14	Wear parts have to be replaced after short operational time.	Cutting tools have to be regrinded after short operational time.	Cutting tools have to be regrinded after a longer operational time.	The maintenance effort is low. Higher maintenance effort is very low even there is a high share of rocks.	Almost no maintenance is required. The wear is very low even there is a high share of rocks.
suitability on very hilly terrain/height guidance/sward care	0.15	Elements for height guidance do not exist. The height guidance is only given by active adjustment of the position.	Soil contact and thus, sward damage occur because of the long length if there are short-wave bumps	Slide shoes support the construction. An adjustment to the soil profile is not always guarenteed if there are short-wave bumps.	The advice is short. Thus, an adaption to short-wave bumps is possible.	The advice adapts to the soil profile very well, both in direction of travel and lateral to the direction of travel. The sward is not damaged during operation.

criteria	weighting factor	scoring points				
		0	1	2	3	4
chopping of cut material	0.09	The cut material is not chopped.	The cut material is chopped sufficiently if the grass is low.	The mowed material is chopped sufficiently.	The mowed material is chopped finely.	The mowed material is chopped very finely.
spreading of cut material	0.11	The cut material is not spread. There is a strong accumulation of the mowed material.	The spreading of cut material is relatively not sufficient if there are especially high and thick plants.	The spreading of cut material is even.	The spreading occurs sufficiently evenly over the whole width, even if there are massive, lignified, lying plants. Under moist conditions clumps are formed.	The spreading is optimal, even if there are massive, lignified, lying plants. Clumps are not formed under moist conditions.
mass	0.09	> 500 kg/m_{AB}	500-400 kg/m_{AB}	300-400 kg/m_{AB}	300-200 kg/m_{AB}	< 200 kg/m_{AB}
safety	0.21	Foreign matters are thrown out uncontrollably. The effective range of the tools is not surrounded by a housing.	Foreign matters can be thrown out uncontrollably.	Foreign matters can be thrown out in limited directions.	Foreign matters are not thrown out. The effective range of the tools is not surrounded by a housing.	Foreign matters are not thrown out. The effective range of the tools is surrounded by a housing.
damage of the fauna	0.04	risk class 3.2-4 high risk	risk class 2.4-3.2 medium to high risk	risk class 1.6-2.4 slight to medium risk	risk class 0.8-1.6 slight risk to no risk	risk class <0.8 no risk

Annexe 4

Mathematical relations of spot detection by a 2D laser scanner (chapter 6.5)

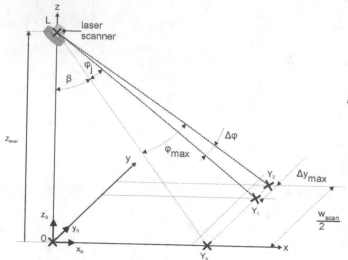

Figure A4.1: Sketch of spot detection by a 2D laser scanner

Approach of calculation via triangles:

$\Delta Y_0 Y_2 L$:   I: $tan\varphi_{max} = \frac{0.5 \cdot w_{scan}}{\overline{LY_0}} \rightarrow \varphi_{max} = arctan\left(\frac{0.5 \cdot w_{scan}}{\overline{LY_0}}\right)$ (A4.1)

$\Delta Y_0 Y_1 L$:   II: $tan(\varphi_{max} - \Delta\varphi) = \frac{0.5 \cdot w_{scan} - \Delta y_{max}}{\overline{LY_0}}$

$\rightarrow \varphi_{max} = arctan\left(\frac{0.5 \cdot w_{scan} - \Delta y_{max}}{\overline{LY_0}}\right) + \Delta\varphi$ (A4.2)

$\Delta O Y_0 L$:   III: $cos(\beta) = \frac{z_{laser}}{\overline{LY_0}} \rightarrow \overline{LY_0} = \frac{z_{laser}}{cos(\beta)}$ (A4.3)

Process of calculation:

I = II: $arctan\left(\frac{0.5 \cdot w_{scan} - \Delta y_{max}}{\overline{LY_0}}\right) + \Delta\varphi = arctan\left(\frac{0.5 \cdot w_{scan}}{\overline{LY_0}}\right)$ (A4.4)

Final result:

+III: $\Delta y_{max\_j} = 0{,}5 \cdot w_{scan} - \underbrace{\frac{z_{laser}}{cos(\beta)} \cdot tan\left[arctan\left(\frac{0{,}5 \cdot w_{scan} \cdot cos(\beta)}{z_{laser}}\right) - j \cdot \Delta\varphi\right]}_{y_j}$ (A4.5)

Annexe 5

Calculation of the kinematics of the steering linkage (chapter 7.1)

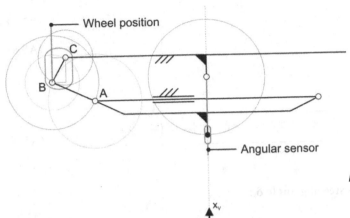

Figure A5.1: Kinematics of the steering linkage

Calculation approach:

I: $$(x_C - x_B)^2 + (y_C - y_B)^2 = \overline{BC}^2 \qquad (A5.1)$$

II: $$(x_B - x_A)^2 + (y_B - y_A)^2 = \overline{AB}^2 \qquad (A5.2)$$

Process of calculation:

I´: $$x_C^2 - 2 \cdot x_C \cdot x_B + x_B^2 + y_C^2 - 2 \cdot y_C \cdot y_B + y_B^2 - \overline{BC}^2 = 0 \qquad (A5.3)$$

II´: $$x_A^2 - 2 \cdot x_A \cdot x_B + x_B^2 + y_A^2 - 2 \cdot y_A \cdot y_B + y_B^2 - \overline{AB}^2 = 0 \qquad (A5.4)$$

I-II: $$(x_C^2 - x_A^2) + (-2 \cdot x_C + 2 \cdot x_A) \cdot x_B + (y_C^2 - y_A^2) \\ + (-2 \cdot y_C + 2 \cdot y_A) \cdot y_B - \overline{BC}^2 + \overline{AB}^2 = 0 \qquad (A5.5)$$

$$\rightarrow x_B = \frac{x_A^2 - x_C^2 + y_A^2 - y_C^2 - (2 \cdot y_A - 2 \cdot y_C) \cdot y_B + \overline{BC}^2 - \overline{AB}^2}{2 \cdot (x_A - x_C)}$$

Substitution:

$$a = x_A{}^2 - x_C{}^2 + y_A{}^2 - y_C{}^2 + \overline{BC}^2 - \overline{AB}^2 \qquad (A5.6)$$

$$b = 2 \cdot (y_A - y_C) \qquad (A5.7)$$

$$c = 2 \cdot (x_A - x_C) \qquad (A5.8)$$

Interim result:

$$y_{B1/2} = \frac{\left(\frac{2 \cdot a \cdot b}{c^2} - \frac{2 \cdot x_A \cdot b}{c} + 2 \cdot y_A\right) \pm \sqrt{\left(\frac{2 \cdot a \cdot b}{c^2} - \frac{2 \cdot x_A \cdot b}{c} + 2 \cdot y_A\right)^2 - 4 \cdot \left(\frac{b^2}{c^2} + 1\right) \cdot \left(\frac{a^2}{c^2} + x_A{}^2 + y_A{}^2 - \overline{AB}^2\right)}}{2 \cdot \left(\frac{b^2}{c^2} + 1\right)} \qquad (A5.9)$$

Relation between y$_B$ and steering angle δ$_s$:

$$\delta_s = arcsin\left(\frac{y_B - y_{B0}}{\overline{BC}}\right) - \delta_{s0} \qquad \text{with angle of offset } \delta_{s0} \qquad (A5.10)$$

$$\delta_s = arcsin\left(\frac{y_B - y_C}{\overline{BC}}\right) - arcsin\left(\frac{y_{B0} - y_C}{\overline{BC}}\right) \qquad (A5.11)$$

$$\rightarrow sin\left(\delta_s + arcsin\left(\frac{y_{B0} - y_C}{\overline{BC}}\right)\right) \cdot \overline{BC} + y_C = y_B \qquad (A5.12)$$

$$\delta_s = arcsin\left(\frac{\frac{\left(\frac{2 \cdot a \cdot b}{c^2} - \frac{2 \cdot x_A \cdot b}{c} + 2 \cdot y_A\right) \pm \sqrt{\left(\frac{2 \cdot a \cdot b}{c^2} - \frac{2 \cdot x_A \cdot b}{c} + 2 \cdot y_A\right)^2 - 4 \cdot \left(\frac{b^2}{c^2} + 1\right) \cdot \left(\frac{a^2}{c^2} + x_A{}^2 + y_A{}^2 - \overline{AB}^2\right)}}{2 \cdot \left(\frac{b^2}{c^2} + 1\right)} - y_C}{\overline{BC}}\right) - arcsin\left(\frac{y_{B0} - y_C}{\overline{BC}}\right) \qquad (A5.13)$$

Detection angle of angular sensor α$_a$:

$$\alpha_a = arctan\left(\frac{\Delta y_A}{x_C - x_A}\right) \qquad (A5.14)$$

Consequence on the turning radius R:

$$\frac{\delta_a}{i} = \frac{L}{R + \frac{b}{2}} \text{ (Source: [73])} \qquad \rightarrow R = \frac{L}{\delta_a} - \frac{b}{2} \qquad (A5.15)$$

Conversions:

Γ:
$$x_C^2 - 2 \cdot x_C \cdot x_B + x_B^2 + y_C^2 - 2 \cdot y_C \cdot y_B + y_B^2 - \overline{BC}^2 = 0 \tag{A5.16}$$

Π:
$$x_A^2 - 2 \cdot x_A \cdot x_B + x_B^2 + y_A^2 - 2 \cdot y_A \cdot y_B + y_B^2 - \overline{AB}^2 = 0 \tag{A5.17}$$

Γ`:
$$x_{B_{1/2}} = \frac{2 \cdot x_C \pm \sqrt{(-2 \cdot x_C)^2 - 4 \cdot 1 \cdot (x_C^2 + x_B^2 + y_C^2 - 2 \cdot y_C \cdot y_B - \overline{BC}^2)}}{2 \cdot 1} \tag{A5.18}$$

Π`:
$$y_{A_{1/2}} = \frac{2 \cdot y_B \pm \sqrt{(-2 \cdot y_B)^2 - 4 \cdot 1 \cdot (x_A^2 + x_B^2 + y_B^2 - 2 \cdot x_A \cdot x_B - \overline{AB}^2)}}{2 \cdot 1} \tag{A5.19}$$

$$\rightarrow \Delta y_A = y_{A_{1/2}} - y_{A_0} \qquad \rightarrow \alpha_a = arctan\left(\frac{\Delta y_A}{x_C - x_A}\right) \tag{A5.20}$$

Annexe 6

Calculation according the displacement Δx_{GPS} of the GPS antenna (chapter 7.1)

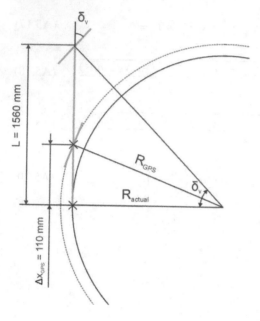
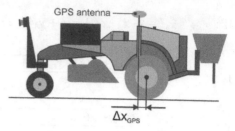

Figure A6.1: Positions of the centre point of the rear axle and the GPS antenna during a circular tour

$$\delta = atan\left(\frac{L}{R_{actual}}\right) \tag{A6.1}$$

$$\delta = atan\left(\frac{L}{\sqrt{R_{GPS}^2 - \Delta x_{GPS}^2}}\right) \tag{A6.2}$$

Calculation of the particular curve radius for the maximum steering angle of the vehicle:

$$R_{GPS} = \sqrt{\left(\frac{L}{tan(\delta)}\right)^2 + \Delta x_{GPS}^2} = \sqrt{\left(\frac{1560\ mm}{tan(50°)}\right)^2 + 110\ mm^2} = 1313\ mm \tag{A6.3}$$

$$R_{actual} = \frac{L}{tan(\delta)} = \frac{1560\ mm}{tan(50°)} = 1308\ mm \tag{A6.4}$$

→ The difference of both is negligible.

Annexe 7

Conversion of velocities (chapter 7.1)

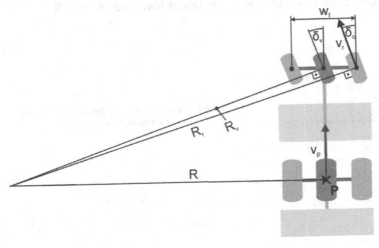

Figure A7.1: Single track model of the machine

Geometric relation of radii and velocity vector:

$$\frac{R_r}{R} = \frac{v_r}{v_P} \qquad \rightarrow v_P = \frac{R}{R_r} \cdot v_r \qquad \text{(A7.1)}$$

Geometric relation of triangles:

$$sin(\delta_r) = \frac{L}{R_r} \qquad \rightarrow R_r = \frac{L}{sin(\delta_r)} \qquad \text{(A7.2)}$$

$$tan(\delta_v) = \frac{L}{R} \qquad \rightarrow R = \frac{L}{tan(\delta_v)} \qquad \text{(A7.3)}$$

Both result finally in the following relationship between v_P and v_r.

$$v_P = \frac{\Delta s_v \cdot v_r}{\Delta s_r} = \frac{R_v}{R_r} \cdot v_r = \frac{sin\delta_r}{tan\delta_v} \cdot v_r \qquad \text{(A7.4)}$$

Annexe 8

Calculation of the time for lifting the mulcher after having finished mulching a spot and until the mulcher must latest lowered in case 2 (chapter 8.3.1)

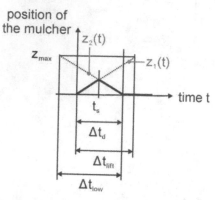

Figure A8.1: Qualitative time behaviour of lifting and lowering the mulcher

Linear equations:

$$z_1(t) = \frac{z_{max}}{\Delta t_{lift}} \cdot t \tag{A8.1}$$

$$z_2(t) = -\frac{z_{max}}{\Delta t_{lift}} \cdot t + C \tag{A8.2}$$

Point of intersection:

$$z_1(t) = z_2(t) \tag{A8.3}$$

$$\frac{z_{max}}{\Delta t_{lift}} \cdot t_s = \frac{z_{max}}{\Delta t_{low}} \cdot t_s + C \rightarrow C = \left(\frac{1}{\Delta t_{lift}} + \frac{1}{\Delta t_{low}} \right) \cdot z_{max} \cdot t_s \tag{A8.4}$$

$$z_2(t) = -\frac{z_{max}}{\Delta t_{low}} \cdot t + t_s \cdot z_{max} \cdot \left(\frac{1}{\Delta t_{lift}} + \frac{1}{\Delta t_{low}} \right) \tag{A8.5}$$

$$z_2(t = t_d) = 0 = -\frac{z_{max}}{\Delta t_{low}} \cdot t_d + t_s \cdot z_{max} \cdot \left(\frac{1}{\Delta t_{lift}} + \frac{1}{\Delta t_{low}} \right) \tag{A8.6}$$

$$0 = -\frac{1}{\Delta t_{low}} \cdot t_d + t_s \cdot \left(\frac{1}{\Delta t_{lift}} + \frac{1}{\Delta t_{low}} \right) \tag{A8.7}$$

$$t_s = \frac{\frac{1}{\Delta t_{low}}}{\left(\frac{1}{\Delta t_{lift}} + \frac{1}{\Delta t_{low}} \right)} \cdot t_d \rightarrow \Delta t^*_{lift} = t_s = \frac{\Delta t_d}{\left(1 + \frac{\Delta t_{low}}{\Delta t_{lift}} \right)} \tag{A8.8}$$

Annexe 9

Calculation of the time t_b depending on the measured fuel consumptions of the machines regarding the effectiveness of selective mulching (chapter 8.3.2)

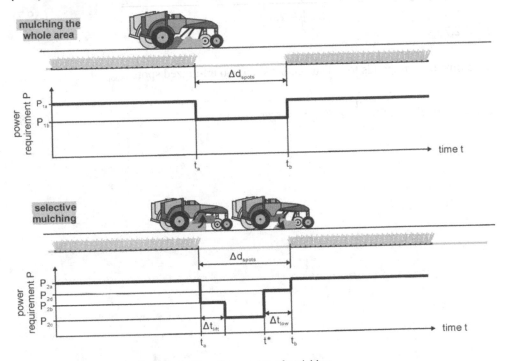

Figure A9.1: Power requirement of different scenarios of mulching

$$\int_{t_a}^{t_b} \sum \dot{V}_{1i} \, dt < \int_{t_a}^{t_b} \sum \dot{V}_{2i} \, dt \tag{A9.1}$$

$$(t_b - t_a) \cdot \dot{V}_{1b} > \Delta t_{lift} \cdot \dot{V}_{2b} + (t^* - t_a - \Delta t_{lift}) \cdot \dot{V}_{2c} + \Delta t_{low} \cdot \dot{V}_{2d} \tag{A9.2}$$

with $t^* = t_b - \Delta t_{low}$ and $t_a = 0 \, s$ (A9.3)

$$t_b \cdot \dot{V}_{1b} > \Delta t_{lift} \cdot \dot{V}_{2b} + (t_b - \Delta t_{low} - \Delta t_{lift}) \cdot \dot{V}_{2c} + \Delta t_{low} \cdot \dot{V}_{2d} \tag{A9.4}$$

$$t_b \cdot (\dot{V}_{1b} - \dot{V}_{2c}) > \Delta t_{lift} \cdot \dot{V}_{2b} - \Delta t_{low} \cdot \dot{V}_{2c} - \Delta t_{lift} \cdot \dot{V}_{2c} + \Delta t_{low} \cdot \dot{V}_{2d} \tag{A9.5}$$

$$t_b > \frac{\Delta t_{lift} \cdot \dot{V}_{2b} - \Delta t_{low} \cdot \dot{V}_{2c} - \Delta t_{lift} \cdot \dot{V}_{2c} + \Delta t_{low} \cdot \dot{V}_{2d}}{(\dot{V}_{1b} - \dot{V}_{2c})} \qquad (A9.6)$$

$$t_b > \frac{700\ ms \cdot 2.6\frac{l}{h} - 1000\ ms \cdot 2.6\frac{l}{h} - 700\ ms \cdot 2.6\frac{l}{h} + 1000\ ms \cdot 4.8\frac{l}{h}}{3.4\frac{l}{h} - 2.6\frac{l}{h}} \qquad (A9.7)$$

$$t_b > 2.75\ s$$

This time t_b corresponds to a distance between two un-grazed spots a_{spots} of

$$a_{spots} > v \cdot t_b = 0.58\ \frac{m}{s} \cdot 2.75\ s = 1.6\ m \qquad (A9.8)$$

Bibliography

[1]   N. Alexandratos and J. Bruinsma: *World agriculture towards 2030/2050 - The 2012 Revision*, Food and Agriculture Organization (FAO), 2012

[2]   Food and Agriculture Organization (FAO), edited by Jelle Bruinsma: World agriculture: towards 2015/2030 - AN FAO PERSPECTIVE, Earthscan Publications Ltd, 2003

[3]   Food and Agriculture Organization (FAO): FAOSTAT (http://faostat3.fao.org) - Domain Code: QL (Livestock Primary); AreaCode: 5000 (World); Element Code: 5318 (Milk Animals); ItemCode: 882 (Milk, whole fresh cow); Year: 1961 - 2013; acccessed 28 Decembre 2015

[4]   Food and Agriculture Organization (FAO): FAOSTAT (http://faostat3.fao.org) - Domain Code: QL (Livestock Primary); AreaCode: 5000 (World); Element Code: 5320 (Producing Animals/Slaughtered); ItemCode: 876 (Meat, cattle); Year: 1961 - 2013; acccessed 28 Decembre 2015

[5]   Food and Agriculture Organization (FAO): FAOSTAT (http://faostat3.fao.org) - Domain Code: QL (Livestock Primary); AreaCode: 5100 (Africa), 5200 (America), 5300 (Asia), 5400 (Europe), 5500 (Oceania); Element Code: 5318 (Milk Animals); ItemCode: 882 (Milk, whole fresh cow); Year: 1961 - 2013; acccessed 28 Decembre 2015

[6]   Food and Agriculture Organization (FAO): FAOSTAT (http://faostat3.fao.org) - DomainCode: EL (Land); AreaCode: 5000 (World); ElementCode: 7208 (% of Agricultural Area); ItemCode: 6655 (Permanent meadows and pastures); Year: 1961 – 2013; acccessed 28 Decembre 2015

[7]   Food and Agriculture Organization (FAO): FAOSTAT (http://faostat3.fao.org) - DomainCode: EL (Land); AreaCode: 5000 (World); ElementCode: 7209 (% of Land Area); ItemCode: 6610 (Agricultural area); Year: 1961 - 2013; acccessed 28 Decembre 2015

[8]   Food and Agriculture Organization (FAO): FAOSTAT (http://faostat3.fao.org) - DomainCode: QL (Livestock Primary); AreaCode: 5100 (Africa), 5200 (America), 5300 (Asia), 5400 (Europe), 5500 (Oceania); ElementCode: 5320 (Producing Animals/Slaughtered); ItemCode: 867 (Meat, cattle); Year: 1961 - 2013; acccessed 28 Decembre 2015

[9]   Eurostat: Agricultural accounts and prices - Statistics Explained, http://ec.europa.eu/eurostat/statistics-explained/index.php/Agricultural_accounts_and _prices#Further_Eurostat_information, acccessed 22 April 2017

© The Editor(s) (if applicable) and The Author(s), under exclusive license to Springer-Verlag GmbH, DE, part of Springer Nature 2020
B. Seiferth, *Development of a system for selective pasture care by an autonomous mobile machine*, Fortschritte Naturstofftechnik,
https://doi.org/10.1007/978-3-662-61655-0

[10]   L. Mannetje: The importance of grazing in temperate grasslands - *Grazing Management - The principles and praktice of grazing, for profit and environmental gain, within temperate grassland systems*, B. G. Society, 2000, pp. 3-13

[11]   Eurostat - Statistics Explained (http://ec.europa.eu/eurostat/): *Agri-environmental indicator - livestock patterns*, European Union (EU), 2012

[12]   Statistisches Bundesamt, *Land- und Forstwirtschaft, Fischerei - Wirtschaftsdünger, Stallhaltung, Weidehaltung - Landwirtschaftszählung/Agrarstrukturerhebung 2010 [Agriculture, forestry, fishing, ... - agricultural census and agricultural structure survey]*, Fachserie 3, Heft 6, Wiesbaden, 2011

[13]   W. Brade: Milchkühe mit hoher Leistung - Vor- und Nachteile der Weidehaltung (Dairy cows with high performance - Advantages and disadvantages of pasture grazing), *Bauernblatt*, pp. 48-49, 2014

[14]   S. Kühl, M. Ermann and A. Spiller: Imageträger Weidegang (Pasture as image carrier), *DLG-Mitteilungen*, no. 4/4014, 2014, pp. 94-97

[15]   H. Bartussek: Die Weidehaltung von Milchkühen aus der Sicht des Tierschutzes (The pasture grazing of dairy cows from the point of view of animal welfare), *Alpenländisches Expertenforum with the topic "Zeitgemäße Weidewirtschaft (Contemporary pasture farming)*, BAL Gumpenstein, Irdning, 1999

[16]   T. Herlitzius, A. Ruckelshausen and K. Weidig, *Mobile Cyber Physical System concept for controlled*, VDI-Bericht 2251 ed., Düsseldorf: VDI Verlag, 2015.

[17]   J. Galler: Grünlandverunkrautung - Ursachen*Vorbeugung*Bekämpfung (Weed infestation of grasslands - reasons*prevention*weed control), Graz, Austria, Leopold Stocker Verlag, 1989

[18]   E. Klapp: Wiesen und Weiden - eine Gründlandlehre (Meadows and pastures - a grassland science), Berlin and Hamburg, Paul Parey, 1971, pp. 245-246

[19]   C. Fox: Grazing techniques for profit: a farmer's perspective, *Grazing Management - The principles and practice of grazing for profit and environmental gain within temperate grassland systems*, B. G. Society, 2000, pp. 221-225

[20]   S. Blackmore, B. Stout, M. Wang and B. Runow: Robotic agriculture - the future of agricultural mechanisation?, *Fifth European Conference on Precision Agriculture (ed. J. Stefford, V.)*, The Netherlands, Wageningen Academic Publishers, 2005

[21]   S. Aigner, G. Egger, G. Gindl and K. Buchgraber: Almen bewirtschaften - Pflege und Management von Almweiden (Cultivating an alp - maintenance and management of alpine pastures), Graz, Austria, Leopold Stocker Verlag, 2003

[22]   A. Niesel, H. Breloer, D. Junker, M. Klärle, B.-H. Lay, D. Münstermann, K. Neumann, A. Steidle-Schwahn, M. Thierme-Hack, J. Thomas and W. Ziegler: Grünflächen-Pflegemanagement - Dynamische Pflege von Grün (Management of green area maintenance - Dynamic care of Green), Stuttgart, Eugen Ulmer, 2006

[23]   Baywa AG: *Ernten Sie Erfolg – mit Landwirtschaftlichen Mischungen von Planterra.(Harvest success - with agricultural mixtures from Planterra.)*, München, 2011

[24]   E. Jedicke, W. Frey, M. Hundsdorfer and E. Steinbach: Praktische Landschaftspflege - Grundlagen und Maßnahmen (Practical landscape conservation - principles and actions), Suttgart, Eugen Ulmer, 1996, p. 224

[25]   K.-H. Kromer and M. Löbbert: *Forschungsberichte Heft Nr. 45*, R. F. Bonn, 1996

[26]   Landwirtschaftlicher Pflanzenbau (Agricultural plant production), München, BLV-Buchverlag, 2014

[27]   Agria Werke GmbH: *Betriebsanleitung (manuals)- Ferngesteuerter Rasenmäher (remote controlled mower) agria 5700 RC hybrid 65 W*, Möckmühl (Germany)

[28]   Dvorak maschine division: *Rotary slope mower - Operation manual Spider ILD01*, Havlickuv Brod, Czech Republic, 2011

[29]   Energreen: *Data sheet - Remote controlled equipment carrier RoboZERO*, Cagnano di Pojana Maggiore (VI), Italy, 2014

[30]   J. Harms and G. Wendl: Automatische Melksystem - Trends, Entwicklungen, Umsetzung (Automatic milking systems - trends, developments, implementation), *39. Viehwirtschaftliche Fachtagung*, Irdning, 2012

[31]   R. Oberschätzl and B. Haidn: *Automatische Fütterungssysteme für Rinder - Technik, Leistung, Planungshinweise (Automatic feeding systems for cattle - Technique, performance, design information)*, Frankfurt am Main: DLG, 2014

[32]   M. Robert and T. Lang: Development of simulation based algorithms for livestock robots, *Landtechnik 68 (4)*, 2013, pp. 278-280

[33]   S. Lefting: Futter nachschieben wie von Geisterhand (feed pushing as if by magic), *topagrar*, No. 11, 2011, pp. R30-31

[34]   DeLaval: Flexible robot removes manure on slats - DeLaval robot scraper RS250
       (product flyer),
       http://www.delaval.ca/ImageVaultFiles/id_209/cf_5/robot_scraper_RS250.pdf,
       acccessed 26 May 2017

[35]   PRECISION MAKERS B.V.: Greenbot - Easy allround machine that functions
       fully independent, http://www.greenbot.nl/greenbot-easy-allround/, acccessed 26
       January 2017

[36]   Lely: LELY VOYAGER - Automatic grazing system - Frontal grazing: the
       innovative way (product brochure),
       http://www.asimo.pl/materialy/download/voyager.pdf, acccessed 26 May 2017

[37]   R. Greenall, E. Warren and M. Warren: A better understandig - Automatic
       Milking, *Integrating Automatic Milking Installations (AMIs) into Grazing
       Systems - Lessons from Australia*, Wageningen, The Netherlands, 2004

[38]   G. Plesch, M. Wittmann and H. Laser: Weide & Melkroboter (Pasture & milking
       robot), 2013, Fachhochschule Südwestfalen (University of applied Sciences),
       https://www4.fh-
       swf.de/media/downloads/fbaw_1/download_1/professoren_1/wittmann/workshop
       /Infoflyer_Empfehlungen_Weide_und_Melkroboter.pdf, acccessed 25 January
       2017

[39]   A. Jukan, X. Masip-Bruin and A. Nina: Smart Computing and Sensing
       Technologies   for   Animal   Welfare: A   Systematic   Review,   2016,
       https://arxiv.org/abs/1609.00627, acccessed 25 January 2017

[40]   Fendt, AGCO GmbH: Presseinformation - Fendt schickt neuen Roboter "Xaver"
       aufs Feld (pressinformation - Fendt sends new robot "Xaver" to the field), 2017,
       https://www.fendt.com/de/fendt-xaver.html, acccessed 19 May 2018

[41]   J. Utz and T. Buchner: *Einzelkornsaateinheit für mobile Agrarroboter - MARS
       Forschungsprojekt (Precision seeding unit for mobile agricultural robots - MARS
       research project),* Düsseldorf, VDI Verlag, 2016

[42]   W. Bangert, A. Kielhorn, F. Rahe, A. Albert, P. Biber, S. Grzonka, S. Haug, A.
       Michaels, D. Mentrup, M. Hänsel, D. Kinski, K. Möller, A. Ruckelshausen, C.
       Scholz, F. Sellmann, W. Strothmann and D. Trautz: Field-Robot-Based
       Agriculture: "RemoteFarming.1" and "BoniRob-Apps", *LAND.TECHNIK –
       AgEng 2013*, Düsseldorf (Germany), 2013

[43]   A. B. Jacobsen, *Crop row navigation for autonomous field robot (Master's
       thesis),* Technical University of Denmark - Department of Electrical Engineering,
       Lyngby, 2015

[44] A. Ruckelshausen: *Imaging und intelligente Sensorsysteme - Schlüsseltechnologien innovativer Landtechnik (Imaging and intelligent sensor systems - key technologies of innovative agricultural engineering), Presentation at the VDI-Seminar Landtechnik*, Freising-Weihenstephan, 2015

[45] J. Hertzberg: Autonome Systeme in der Landwirtschaft (Autonomous systems in agriculture), KTBL-Tagung 2014 in Potsdam *"Vernetzte Landtechnik - Nutzen für die Betriebsführung (Networked agricultural technology - benefit for farm management)"*, Kuratorium für Technik und Bauwesen in der Landwirtschaft e.V. (KTBL), Darmstadt, 2014, pp. 57-63

[46] N. Shalal, T. Low, C. McCarthy and N. Hancock: A review of autonomous navigation systems in agricultural environments, *Innovative Agricultural Technologies for a Sustainable Future*, Barton, Australia, 2013

[47] M. Kise, Q. Zhang and N. Noguchi: An obstacle identification algortihm for a laser range finder-based obstacle detector, *Transactions of the ASAE Vol.48(3)*, pp. 1269-1278, 2005

[48] *Agricultural machinery and tractors - Safety of highly automated machinery*, International Organisation fo Standardisation (ISO), 2013

[49] T. Herlitzius, A. Grosa, M. Henke, K. Krzywinski, F. Pahner and M. Klingner: *Concept Study of a Modular and Scalable Self - Propelled*, VDI-Bericht 2193, Düsseldorf, VDI Verlag, 2013

[50] A. Steinwidder and W. Starz: Gras dich fit! - Weidewirtschaft erfolgreich umsetzen (Graze fit - Implement pasture farming successfully), Graz, Leopold Stocker, 2015

[51] C. Schiess-Bühler, R. Frick, B. Stäheli and R. Furi: *Erntetechnik und Artenvielfalt in Wiesen (Harvesting technology and biodiversity in grasslands)*, AGRIDEA, Lindau, Lausanne, 2011

[52] Leica Geosystems AG: *GPS Basics - Introduction to GPS (Global Positioning System)*, Heerbrugg, Switzerland, 1999

[53] J. Hüter and U. Klöble: *Precision Farming in der Praxis - Technik und Anwendungsmöglichkeiten (Precision farming in practice - technique and possible applications)*, Kuratorium für Technik und Bauwesen in der Landwirtschaft e.V. (KTBL), Darmstadt, 2007

[54] H. Durrant-Whyte: *A Critical Review of the State-of-the-Art in Autonomous Land Vehicle Systems and Technology*, Sandia National Laboratories, 2001

[55] M. Juha Hyyppä: State of the Art in Laser Scanning, *Photogrammetric Week '11*, Berlin, Offenbach, 2011

[56] C. Wellington, J. Campoy, L. Khot and R. Ehsani: Orchard Tree Modeling for Advanced Sprayer Cotnrol and Automatic Tree Inventory, *IEEE/RSJ International Conference on Intelligent Robots and Systems (IROS) Workshop on Agricultural Robotics*, 2012

[57] A. Linz, D. Brunner, J. Fehrmann, M. Grimsel, T. Herlitzius, R. kKeicher, A. Ruckelshausen, H.-P. Schwarz and E. Wunder: "elWObot" - A Diesel-Electic Autonomous Sensor Conrolled Service-Robot for Orchards and Vineyards, *4th International Conference on Machine Control & Guidance*, Braunschweig, 2014

[58] P. Thanpattranon, T. Ahamed and T. Takigawa: Navigation of an Autonomous Tractor for a Row-Type Tree Plantation Using a Laser Range Finder— Development of a Point-to-Go Algorithm, *Robotics (ISSN 2218-6581)*, 2015, pp. 341-364

[59] K.-H. Lee and R. Ehsani: Comparison of two 2D laser scanners for sensing object distances, shapes, and surface patterns, *Computers and Electronics in Agriculture (Vol. 60)*, pp. 250-262, 2008

[60] Z. Doerr, A. Mohsenimanesh, C. Laguë and N. McLaughlin: Application of the LIDAR technology for obstacle detection during the operation of agricultural vehicles, *Canadian Biosystems Engineering Volume 55*, 2013

[61] N. Pfeifer, B. Höfle, C. Briese, M. Rutzinger and A. Haring: Analysis of the backscatterd energy in terrestrial laser scanning data, *The International Archives of the Photogrammetry, Remote Sensing and Spatial Information Sciences 37*, 2008

[62] J. Billingsley and M. Schoenfisch: The successful development of a vision guidance system for agriculture, *Computers and Electronics in Agriculture*, Nr. Volume 16, Issue 2, 1997, pp. 147-163

[63] E. Benson, J. Reid and Q. Zhang: Machine Vision Based Steering System for Agricultural Combines, *The Society for engineering in agricultural, food, an biological systems (ASAE) Annual Internation Meeting*, Sacramento, California, USA, 2001

[64] P. Fleischmann and K. Berns: A Stereo Vision Based Obstacle Detection, *Field and Service Robotics*, Springer, 2015, pp. 217-231

[65] M. Holpp and L. Dürr: Automatisches Lenksystem für Traktoren im Obstbau (Automatic guidance system for tractors in fruit farming), *Agrarforschung Schweiz 15(2)*, 2008, pp. 108-113

[66]   J. A. Parish, J. D. Rivera, H. T. Boland and R. Lemus: *Beef Cattle Grazing Management*, Mississippi State University, 2010

[67]   International Committee for Animal Recording (ICAR): *Results for the years 2012 - 2013 - Yearly enquiry on the situation of cow milk recording in ICAR member countries*, ICAR, 2014

[68]   W. Dietl and J. Lehmann: Ökologischer Wiesenbau - Nachhaltige Bewirtschaftung von Wiesen und Weiden (Ecological grassland farming - sustainable farming of meadows and pastures), Leopoldsdorf, Austria, Österreichischer Agrarverlag, 2004

[69]   G. Fröhlich, K.-H. Bröker, H. Link, G. Rödel, F. Wendling and G. Wendl: Computer-Controlled Plot Fertilizer Spreaders for Precision Experiments, *Landtechnik 3/2007*, 2007, pp. 150-151

[70]   A. Ruckelshausen, R. Klose, A. Linz, M. Gebben and S. Brinkmann: Intelligent Sensor System for Autonomous Field Robots, *Bornimer Agrartechnische Berichte*, Nr. Heft 62, 2008, pp. 67-75

[71]   O. C. Jr. Barawid, A. Mizushima, K. Ishii and N. Noguchi: Development of an Autonomous Navigation System using a Two-dimensional Laser Scanner in an Orchard Application, *Biosystems Engineering*, Volume 96, Issue 2, 2007, pp. 139-149

[72]   J. R. Rosell, J. Llorens, R. Sanz, J. Arnó, M. Ribes-Dasi, J. Masip, A. Escolà, F. Camp, F. Solanelles, F. Gràcia, E. Gil, L. Val, S. Planas and J. Palacín: Obtaining the three-dimensional structure of tree orchards from remote 2D terrestrial LIDAR scanning, *Agricutlural and Forest Meteorology*, Volume 149, Issue 9, 2009, pp. 1505-1515

[73]   Pepperl&Fuchs: *MANUAL - 2-D Laser Scanner - OMD10M-R2000-B23, OMD30M-R2000-B23*, Mannheim, Germany, 2015

[74]   J. Hertzberg, K. Lingemann and A. Nüchter: Mobile Roboter, Berlin Heidelberg, Springer Verlag, 2012

[75]   National Instruments (NI): *User Guide and Specifications - NI myRIO-1900*, 2013

[76]   Stratom, Inc.: *X-CAN myRIO Hardware - User Guide and Specifications*, Boulder, USA, 2015

[77]   C. Cariou, Z. Gobor, B. Seiferth and M. Berducat: Mobile robot trajectory planning under kinematic and dynamic constraints for partial and full field coverage, *Journal of Field Robotics*, 2017

[78]   C. Cariou, R. Lenain, B. Thuilot and M. Berducat: Automatic guidance of a four-wheel-steering mobile robot for accurate field operations, *Journal of Field Robotics,* No. 26(6/7), 2009, pp. 504-518

[79]   B. Seiferth, S. Thurner, A. Gain and Z. Gobor: Entwicklung eines intelligenten Weidemanagementsystems mit moderner Informations- und Kommunikations-technologie (Development of an intelligent pasture management system with modern Information and Communication Technology (ICT)), *Bau, Technik und Umwelt in der landwirtschaftlichen Nutztierhaltung (BTU),* Stuttgart, 2017

[80]   YANMAR: *DIESELMOTOREN - TMV-Serie - Max. Leistung - 13,4~83,8hp (Diesel engines - TMV-series),* 1332 BS Almere-de Vaart, The Netherlands, 2015

[81]   M. Lösch: Faunaschonende Flächenpflegetechnik - Entwicklung und Beurteilung (Fauna careful area maintenance technique - development and evaluation), Fortschrittberichte VDI, Reihe 14 Landtechnik/Lebensmitteltechnik No. 81, Düsseldorf, VDI Verlag GmbH, 1997

[82]   Arbeitsgemeinschaft Deutscher Rinderzüchter e.V.: ADR - Kontrolldichte in ICAR-Mitgliedsorganisationen,   http://www.adr-web.de/kontrolldichte_in_icarmitglied sorganisationen.html, acccessed 30 March 2015

[83]   W. Korth: *Vorlesung für Master Geoinformation,* Hochschule Beuth, 2009

[84]   S.P. Drake: DSTO Electronics and Surveillance Research Laboratory, *Surveillance Systems Division,* Edinburgh, Commonwealth of Australia, 2002

[85]   Energreen SRL: Spare parts catalogue - ROBOZERO, Cagnano di Pojana Maggiore, Italy, 2015

[86]   C. Cariou, Z. Gobor, B. Seiferth and M. Berducat: Mobile Robot Trajectory Planning Under Kinematic and Dynamic Constraints for Partial and Full Field Coverage, *Journal of Field Robotics,* May 2017

Printed in the United States
By Bookmasters